T0291031

Flyrock in Surface Mining

This book provides a comprehensive understanding of historical and recent research, with a critical review of several aspects of the flyrock phenomenon, along with the classification of pertinent literature. This puts flyrock into proper perspective and develops a comprehensive regime for flyrock prediction and control. It also addresses the blast danger zone demarcation based on scientific understanding in comparison to the consequence-based approach supported by pertinent case studies.

Features:

- Discusses exclusive material on flyrock in surface mining.
- Presents comprehensive and critical review of the flyrock phenomenon.
- Reviews prediction and control mechanisms in vogue with scientific and risk-based definitions of blast danger zone.
- Provides new insights into the flyrock definitions, prediction, and prevention along with the research approach to the problem.
- Includes Indian case studies and summarizes global data available in the published domain.

This book is aimed at researchers and graduate students in mining and civil engineering, engineering geology, and blasting.

Flyrock in Surface Mining

Origin, Prediction, and Control

Autar K. Raina

CRC Press is an imprint of the
Taylor & Francis Group, an **informa** business

First edition published 2024
by CRC Press
6000 Broken Sound Parkway NW, Suite 300, Boca Raton, FL 33487–2742

and by CRC Press
4 Park Square, Milton Park, Abingdon, Oxon, OX14 4RN

CRC Press is an imprint of Taylor & Francis Group, LLC

ISBN: 9781032356112 (HB)
ISBN: 9781032356129 (PB)
ISBN: 9781003327653 (EB)

DOI: 10.1201/9781003327653

Typeset in Times
by Apex CoVantage, LLC

Dedication

Dedicated to Dada (Late Dr. Ashoke Chakraborty), who pushed me to limits; Prof. B.B. Dhar, Babosh and Maa, Bhaiya and Bhabhi—who saw me through thick and thin.

An accident is a chain of events . . .

Anon.

"Improvements are cumulative, then any improvements that we make in our blast planning and executing assuredly will enhance the Quality of our blasting safety. We can rapidly advance our improvement by benchmarking, emulating the best practices. It is essential for the leadership of our explosives industry to underscore the crucial importance of continued blasting safety training for supervisors overseeing blasting operations, for blasters and for the personnel working in blasting crews. Alone, safety training is a paper tiger. We must instill in every blasting person a safe attitude, the sine qua non necessary to achieve the Quality performance standard for blasting safety, zero accidents."

—Brulia (1993)

Author's Take on Flyrock

When things are perfectly engineered the system behaves, but still falters for uncertainties . . . as a rule . . . even if the mine-mill fragmentation system is fully engineered, the uncertainties in the rockmass, human intervention, and explosives provide enough reasons for a fragment to shoot out.

To me flyrock is far more dangerous than ground vibrations. It can damage, injure, and even kill. I am still at loss to understand why people have spent a fortune in ground vibration predictions.

Contents

About the Author

Autar K. Raina is at present working as Chief Scientist in the Mining Technology Department at Nagpur Research Center of CSIR-Central Institute of Mining and Fuel Research, India. He is also Professor at the Academy of Excellence in Scientific and Innovative Research (AcSIR), India. A gold medalist in Post Graduate studies, he has a Ph.D. in Mining Engineering from IIT-ISM, Dhanbad and M.Sc., M.Phil., and Ph.D. in Geology from the University of Jammu, India, to his credit. He is a recipient of the coveted National Mineral (now Geosciences) Award from the Ministry of Mines, Government of India, in addition to several other awards, certificates of merit, and appreciations from the industry for different contributions to rock blasting and use of IT in Mining. With a research experience of more than 25 years in several aspects of blasting like fragmentation, human response to blasting, flyrock, and ground vibrations, he has published more than 150 papers in national and international journals with 16 papers on the subject of flyrock. He has several books, book chapters, and edited chapters on rock blasting to his credit and is on the reviewer panel of several international journals of repute. He has conducted research on several aspects of blasting and selection methodology of tunnel boring machines through several government-funded research projects. He has guided several students for their Ph.D., M.S., and B.E. dissertations. Fragalyst—a digital image analysis software for blast fragmentation measurement—has been consistently upgraded by him for the advantage of mining industry along with other IT apps for rockmass classification and blast design like Blast Design and Database Analysis System (BD2MAS) wherein he has several patents and copyrights to his credit.

Preface

In modern times of internet, ever-growing information, and its flow, it is practically difficult to compile any document that qualifies as a book. The daily increasing scenario of citations makes the task more difficult. Despite this, there are references that qualify as classical works and there are just others which receive little attention or have little content. At the same time, one cannot be oblivious of yesterday's works that might not have caught the attention of the authors but have significant content. Despite this, there are certain questions in science that are pertinent and require a sustained introspection. The author is aware of the fact and the topic of the book chosen is akin to present-day surface mining.

Blasting is important and rather a dominant method of rock excavation, even today. There are several outcomes of blast into which the energy provided by the explosives on detonation gets split up. The major portion of this energy is consumed by ground vibrations and air overpressure. Little energy is used to fragment and heave the rock broken under the influence of explosive pressures. Least energy is however, in some cases, manifested in the form of rock fragments travelling beyond planned distances. Such fragments, known as flyrock, have gained importance after the author of this book provided new insights into the phenomenon with several instances to model the behaviour.

Flyrock has received little attention until 2014 and has recorded a significant growth in its prediction and control thereafter. Since flyrock has a potential to cause fatalities, it should have received significant attention, but the facts are contrary. Vibrations and air overpressure have been dealt with by researchers in detail, though it can only damage the structures and has a nuisance value. The difficulty in acquisition of sufficient data, reporting constrained by the known statutory consequences, occurrence by chance, and difficulty in defining the predictive regime of flyrock can be some of the reasons for the bias, if not otherwise. The multiple mechanics involved in the generation, launch, travel, and landing of a flyrock are additional complications that face the researchers.

The prediction of flyrock is still a herculean task as one cannot produce flyrock for testing, and even if this is tried, there is no guarantee that flyrock will occur in a test. This specifically eludes the researchers and makes it very difficult to predict the same. Unlike fragmentation, throw, and vibrations from blasting, flyrock is an uncertain event. Also, if a flyrock is generated, whether it will travel in a direction of concern and hit any object or person and what could be the level of damage are all uncertain quantities. So, the phenomenon of flyrock has several associated probabilities that not only makes the task of flyrock event and its travel distance difficult to predict but places constraints on defining the blast danger zone around a proposed blast. This has resulted in many regulations that are just flyrock event based and present little scientific explanation. In addition, the consequences are extremely difficult to quantify as lot of subjectivity is involved, particularly if flyrock hits a person. Thus, the consequences are difficult to quantify. However, the probabilities and consequences can together be framed into a network to provide an insight into the risk involved due to flyrock. The method has an advantage that it can allow a dynamic blast danger zone, in contrast to the static zones, about the mines, and in turn allow mining of precious minerals, very close to the habitats. Accordingly, this book has been compiled with focus on literature and its review and to provide means and define terms that can be used by researchers to control the flyrock. In brief, the book is addressed to the students and planners for poking into the issue rather than providing a holistic solution to the problem.

I hope that students and researchers along with field engineers and legislators, who are interested in blasting with a keen interest in the flyrock, will find the book a useful one. I am expecting response on the work, different new propositions and concepts, from seekers of the subject and critics.

The approval of Director, CSIR-Central Institute of Mining and Fuel Research, India, and the financial aid in the form of grant for a research project from the Ministry of Mines, Government of India, are acknowledged by the author. This book would not be possible without active support of my office and its staff, particularly at Nagpur Research Center, both scientific and technical. The help rendered by them is praiseworthy. Umpteen number of people from various mining sectors in India helped in data acquisition and provided all sorts of support that is worth mentioning.

My gratitude to Dr. Chakraborty who introduced me to the topic and facilitated several studies. Prof. VMSR Murthy, my Ph.D. guide at IIT-ISM, Dhanbad, critically examined several of my works that proved valuable in shaping this book. Prof. Agne Rustan provided several critical inputs and encouragement. Dr. Saša Stojadinović, Prof. R.N. Gupta, Dr. P. Pal Roy, Dr. A. Sinha, Prof. Arvind K. Mishra, Dr. A.K. Soni, Prof. P. Rai, Dr. Bhanwar, Dr. Suresh Sharma, and Dr. K. Ram Chandar deserve thanks for their contributions. Bhushan, Shantanu, Late Arup, Pawan, Harsha, Rohan, and many of my associates deserve mention for their active help. Anand, Rishi, Dr. Narayan, and many associated fellows deserve special mention for their last moment help. Kusum, Baijee, Acharya Motilal, Jaya maasi, and Prof. Maharaj Pandit deserve special mention for their contributions in my career. Last but not least, my ardhangini Rimple, daughter Medhu, and son Muppu supported me through sharing the time that belonged to them.

Symbols

Δ_t	length of impulse time (s) of flyrock
Δp_τ	adjustment for pressure in Pa for burden distance
A	area of flyrock fragment in m^2
B	burden (m)
B_{ob}	optimum breakage burden
B_c	critical burden (m)
b_d	drag factor (dimensionless)
b_{sd}	specific drilling ($1/m^2$)
B_{sd}	scaled depth of burial of explosive (metric system)
$C(E)$	consequence of an event
c_d	velocity of detonation (m/s)
c_{dc}	confined velocity of detonation (m/s)
c_{di}	ideal velocity of detonation (m/s)
C_f	correction factor (used in throw to account for flyrock)
c_{pi}	p-wave velocity or longitudinal wave velocity (m/s), measured *in situ*
c_{Si}	s-wave velocity or transverse wave velocity (m/s), measured *in situ*
C_x, C_D	drag coefficient (dimensionless)
d	diameter of blasthole (m)
d_c	diameter of cartridge (m)
d_e	diameter of explosive (m)
E	modulus of elasticity or Young's modulus (Pa or GPa)
E'	Gurney's constant
E_{lq}	linear energy of the explosive charge in $\times 10^3$ J/m
f	target impact frequency (impact/year)
f_p	pattern footage
g	acceleration due to gravity in m/s^2
H_1	maximum height (m) along a distance R_h
H_2	maximum height (m) along a distance R_t
H_b	height of bench (m)
H_i	the height of rise in m at any given horizontal distance measured relative to the original elevation of the fragment
I_{BI}	Blasting Index (fragmentation oriented)
	j, k_1, k_2, and k_3 proportionality constants
J_{fr}	joint frequency rating
k_{50}	mean fragment size
k_c	characteristic fragment size
k_f	size of flyrock (m)
k_{opt}	optimum fragment size
K_v	velocity coefficient
k_x	size of the fragment (nominal diameter, m)
L	length of flyrock fragment (m)
l_{bh}	length of borehole or blasthole (m)

L_f	shape factor of flyrock
l_o	backbreak (depth) (m)
l_q	length of charge (m)
l_s	length of stemming (m)
l_{sd}	length of stemming between deck charges (m)
l_{sub}	subdrilling length
m	mass of flyrock fragment (g, kg)
M_e	mass of explosive in one blasthole (kg)
m_l	total mass of material per unit of length (kg/m)
M_r	mass of rock blasted by one blasthole ($\times 10^3$ kg)
n	slope of a function
N_b	total number of blasts per year
N_{ff}	number of free faces
O_j	joint orientation (°)
p	pressure (Pa)
$P(E)$	probability of a flyrock event
$P(R)$	probability of wild flyrock travelling the target distance
$P(T)$	probability of wild flyrock travelling on an impact trajectory
$P(T_e)$	probability of target exposure
p_a	probability of a person being hit by a flyrock
p_b	pressure acting on the flyrock fragment at escape (Pascal)
p_{bc}	near field pressure (Pa or MPa) from blasthole measured in rock at a distance from blasthole
Q	explosive weight (kg)
q	specific charge or old powder factor (kg/m^3 of rock)
q_a	charge factor or load (kg/m^2)
Q_h	mass of explosive charge (kg) in a blasthole (also called as charge per hole)
q_l	linear charge concentration (kg/m)
R	distance from blasthole or blast site to a measuring point (m)
R_0	the curve characteristic
r^2	coefficient of determination (dimensionless)
R_{dc}	decoupling ratio (dimensionless)
R_e	distance of excess throw or length (m) (includes displacement of much smaller number of fragments)
R_f	maximum horizontal flyrock distance (m)
R_{fpt}	maximum horizontal flyrock distance in pressure–time method (m)
R_h	distance travelled by the flyrock along a horizontal line at the original elevation of the rock on the face (m)
R_h	the horizontal distance along the trajectory in m
R_m	distance of throw or throw length (m) of considerable amount of fragmented rock from blast face (m)
R_n	Reynolds number (dimensionless)
R_{obj}	distance of object of concern from the blast site (m)
R_{opt}	optimum throw (m), throw distance for efficient loading of muck by equipment (m)

R_{perm}	permissible or acceptable distance of flyrock (m)
R_t	total distance travelled by a fragment ejected from the blast accounting for its heights above the pit floor (m)
R_T	throw of blasted rockmass (m)
S	spacing between blastholes (m)
S_a	visible surface of a person (for flyrock impact)
S_j	rock joint spacing (m)
t	time (s)
t_{HH}	delay between two blastholes
T_r	threat ratio
t_{RR}	delay between two rows in a blast
v	velocity
V	volume (m^3)
v_0	initial or exit velocity of flyrock fragment (m/s)
V_h	volume of blasthole
W	width of flyrock fragment (m)
W_b	width of bench
W_j	energy required to crush unit weight of rock (J)
W_r	energy absorbed to fragment a unit weight of rock (J)
W_s	seismic energy generated per unit weight of explosive (J)
Z	acoustic impedance of material ($kg/m^2/s$)
Z_e	acoustic impedance of explosive ($kg/m^2/s$)
Z_r	acoustic impedance of rock ($kg/m^2/s$) measured in situ
θ	launch angle of flyrock (°)
π	3.14159
ρ_e	density of explosive before charging (kg/m^3) or g/cm^3
ρ_{ee}	effective in-hole density of explosive (kg/m^3)
ρ_r	density of rock (kg/m^3)
σ_c	uniaxial compressive strength (MPa) (estimated from Schmidt hammer)
Φ	angle of breakage (°)
α	muck angle (°)
α_{bi}	blasthole angle (°)
α_{opt}	optimum muck angle (°)
η_a	gas viscosity coefficient under movement
μ	Poisson's ratio
μ_{air}	air drag (m/s^2)—negative
ρ_{fluid}	the fluid mass density kg/m^3

Acronyms

ACO	ant colony algorithm
AIME	American Institute of Mining, Metallurgical, and Petroleum Engineers
ANFIS	adaptive neuro fuzzy inference system
ANFO	ammonium nitrate and fuel oil
ANN	artificial neural networking
ANN-ADHS	ANN coupled with adaptive dynamical harmony search
ANN-HS	hybrid ANN models coupled by harmony search
	AusIMMAustralasian Institute of Mining and Metallurgy
BBO	biogeography-based optimization
BDZ	blast danger zone
BN	Bayesian network
BRT	boosted regression tree
CA	cultural algorithm
CART	classification and regression tree
CFR	Code of Federal Regulation (USA)
CIM	Canadian Institute of Mining, Metallurgy and Petroleum
COA	cuckoo optimization algorithm
DA	dimensional analysis algorithm
DE	differential evaluation
DF	detonating fuse
DGMS	Directorate General of Mines Safety (India)
DNN	deep neural network
DP	drilling pattern (Rectangular or staggered)
DT	decision tree
DTH	down-the-hole hammer (drills)
ELM	extreme learning machine
EMLM/BN	ensemble machine learning method/Bayesian network, algorithm
ET	explosive type
FBPNN	feedforward back propagation neural network
FCM	fuzzy cognitive map
FFA	firefly algorithm
FoS	factor of safety
FRAGBLAST	shortened form for International Symposium on Fragmentation by Blasting
FRES	fuzzy rock engineering system
Fuzzy	fuzzy logic
GA	genetic algorithm
GDP	gross domestic product
GEP	genetic expression programming
GOA	grasshopper optimization algorithm

GP	genetic programming
GP/GEP	gene expression programming
GSI	Geological Strength Index
GWO	grey wolf optimization
ICA-ANN	imperialist competitive algorithm
ICGCM	International Conference on Ground Control in Mining
ILO	International Labour Organization
IOM3	Institute of Materials, Mineral and Mining—accessed through ISEE OneMine portal (members only portal)
ISEE	International Society of Explosive Engineers
ISEE ODB	International Society of Explosive Engineers Online Database
ISRM	International Society for Rock Mechanics and Rock Engineering
KELM	kernel extreme learning machine
LMR	linear multiple regression
LOO	leave one out cross-validation method
LS-SVM	least squares support vector machines
LWLR	local weighted linear regression
MARS	multivariate adaptive regression splines
MAS	motion analysis software
MHA	metaheuristic algorithm
ML	machine learning
MLR	multiple linear regression analysis
MMFS	mine-mill fragmentation system
MRMR	mining rockmass rating
ms	millisecond when used to classify detonators and delay periods
MSWA	multichannel seismic wave analyser
MWD	measure while drilling
NeSt	non-electric shock tube initiation device
NG-ANN	neurogenetic artificial neural network
NIOSH	National Institute for Occupational Safety and Health
OC	objects of concern
OR-ELM	outlier robust-ELM
PCA	principal component analysis
PCR	principal component regression
PETN	pentaerythritol trinitrate
PSO	particle swarm optimization
RF	random forest
RFNN	recurrent fuzzy neural network
RMSE	root mean square error
RQD	rock quality designation
RSA	response surface analysis
RSM	response surface methodology
SAG	semi-autogenous
SAIMM	Southern African Institute of Mining and Metallurgy
SEM	probabilistic structural equation model
SME	Society for Mining Metallurgy and Exploration

SVM	support vector machine
SVR	support vector regression
TOPSIS	technique for order preference by similarity to ideal solution
USBM	United States Bureau of Mines
WOA	whale optimization algorithm
XRD	X-ray diffraction

1 Introduction

Mining that consumes most of the explosives to date has a tremendous role in society as it not only contributes to the world GDP, but also forms the core of the industrial revolution, as metals and minerals are also produced from mines. No sphere of life on this planet is untouched by mining, including IT, as the basic hardware used in electronics has inking of minerals.

Excavations are places where rockmass is removed for a variety of purposes. In some cases of excavations, only soil or soft rock is encountered that can be removed by mechanical devices like excavators. However, in many cases of excavations, hard rock is met, which needs to be broken and dislodged, before it can be lifted, loaded, and transported. If explosives are used to break such rock to facilitate easy loading and hauling, it is termed as blasting. Hence, blasting is a method of rock breakage used in different types of excavations and is a well-established technique that has been used since times immemorial. Blasting is the core of the mining activity world over.

The objectives of blasting in the diverse projects vary and depend on the final desired outcome of the planned application. In civil structures, it is used to create space for hosting surface or underground structures like foundations, tunnels, defence installations, nuclear power plants, nuclear waste repositories, and underground space creation, in addition to the production of construction material and dimensional rocks. However, the objective of blasting in mines, which produce minerals, is to fragment the rock to achieve a desirable size in tune with the system requirements and at the same time throw and heave the broken rock into a particular profile which is economically feasible for lifting or loading, transportation or hauling, and further breaking by mechanical means to achieve the desired size of fragments. In addition, there are a multitude of applications of blasting, most of which have been compiled in Figure 1.1. Although mechanical methods of excavation are gaining pace with the introduction of efficient tunnel boring machines and roadheaders, blasting continues to be the major method of rock breakage, considering its comparative economical advantage over other methods. It will not be out of place to mention that blasting, which encompasses creation of space or production of minerals, is part of our daily life.

For a common man, blasting entails explosive, threat, and danger to life. Terrorism is also deeply connected with explosives, which for the sake of human interest has not been described further. However, for a blasting engineer, the word "blasting" denotes a complex process of breakage of rockmass through interactions of products of detonating explosive with rockmass for its removal aimed to benefit the humanity and of course, production-related profitability. Accordingly, our focus will be on mining as mines consume the maximum quantities of explosives. The explosive market is estimated to grow from $18,000 million in 2021 to $22,000 million by 2028.[1] Deployment of such volume of explosive in mining also means consequent high-probability dangers associated with blasting.

DOI: 10.1201/9781003327653-1

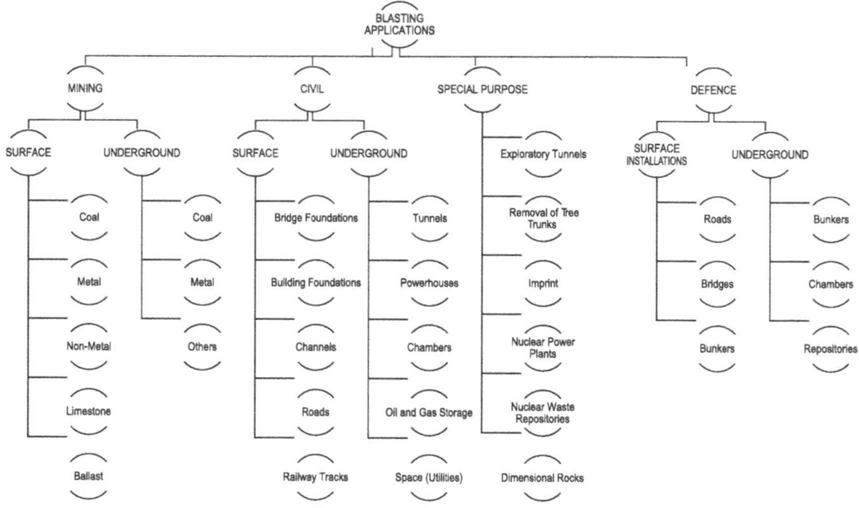

FIGURE 1.1 Broad spectrum of application of blasting in various types of projects.

Blasting in mines through the science of rock breakage has progressed significantly in recent years. With the advent of supercomputing, advanced numerical methods, high-fidelity sensors and high-speed data acquisition systems, interdisciplinary studies, and artificial intelligence methods, the findings related to explosive-induced breakage of rocks have yielded tangible results for field deployment. To drive the topic in hand, it is imperative to have a broad idea of what is being discussed. A blast design props up even at the stages of feasibility and detailed project report formulations for a mining project or even civil works of varied types. The requirement of annual production volume defines the daily production demand in mines and as such, the volume of rock that needs to be blasted at a time. This defines several other components like the drill diameter, the shovel, and the hauler capacities and hence the blast design, as will be explained further. Blast design once tested and accepted in actual conditions transforms into a production pattern.

1.1 BLASTING BASICS

Blasting, or chemical excavation method, involves application of explosive energy to fragment and throw rock. Bhandari (1997) provided a complete information on the explosive energy partitioning and its utilization by productive and unproductive results of blasting. It is believed that only a fraction of energy, supplied to the rock by the explosive, is utilized in fragmenting and throwing the rock to a distance, and most of this energy is transformed into ground vibrations and air overpressure. It may be pointed out that the energetics of a blast is a very complicated subject owing to high speed of explosion in a blast, instant release of energy, and its rapid dissipation. However, there are cases which document the components of the energy entering different domains. An energy transfer efficiency test called the "cylinder expansion test" introduced by Ouchterlony et al. (2004) claims to quantify the partitioning of explosive energy into various components and provide the details of the energy transmission and conversion into seismic, kinetic, and fragmentation energies.

Sanchidrián et al. (2007) concluded that 2–6% of the total available energy is expended in the form of fragmentation, 1–3% for the seismic energy, and 3–21% for the kinetic energy. They added that for a confined blasthole, the seismic energy was 9% of the heat of explosion. Calnan (2015) claimed to account for 73% of the total explosive energy available in a blasthole that included energies for borehole chambering (13%), rotational kinetic energy (25%), translational kinetic energy (5%), and air overpressure (28%) and concluded that borehole chambering, heave, and air blast are the largest energy components in a blast. However, Comeau (2019) argued that major energy may be consumed in crushing to generate particles of <1 mm size and hence estimation of quantity of fines generated is important while providing fragment size distributions during measurement.

One important assertion about blasting is that explosives used in breakage do not know the rockmass and in turn rockmass does not know the explosives. The interaction starts when an explosion takes place in a blasthole. The enormous amount of energy in the range of 10–15 GPa, confined in a blasthole, on release, starts the talk with the rockmass. The huge blasthole pressures that are confined can be held in a blasthole over a very short span of time ranging from few microseconds to few milliseconds. During this process, two major mechanisms come into play.

1. The shock due to sudden release of energy or simply the detonation pressures interact with the rockmass being loaded.
2. The borehole pressure due to highly confined gases try to escape through least resistance paths in multiple ways.

The shock is believed to create fractures in the rockmass and consumes maximum pressure generated by the explosive (Cunningham, 2006) and creates a zone of breakage zone about the blasthole modelled by Esen et al. (2003). The blasthole pressure gets activated simultaneously or with a delay of fraction of second which is believed to expand the existing or newly formed cracks and produce further fragmentation and displaces the fragmented rock to a distance forming a muckpile. The phenomenon is discussed in detail by Mortazavi and Katsabanis (2000) and Sim et al. (2017). A complete description of the blasting fracture mechanics, the theoretical foundations, and its application in rock breakage can be found in Zhang (2016). Different recent studies have focused on impact of various factors like role of initiation point (Long et al., 2013), rockmass discontinuity orientation and their dip on process of burden breakage (Ash, 1973; Mortazavi & Katsabanis, 2000), effect of in situ stresses on rock breakage (Yi et al., 2018), and rock-explosive interactions (Raina & Trivedi, 2019).

Irrespective of the said facts, it is believed that a very small fraction to the tune of 1% of the explosive energy of a blast may propel rock fragments to undesired distances and can be dangerous (Berta, 1990). However, none of the recent research has mentioned the energy component transformed to flyrock, probably because it is not a regular outcome of a blast and is restricted to poor design, human factors, special conditions of rockmass, or malfunctioning of the pyrotechnic delay elements.

Such rock fragments emanating from a blast, travelling beyond expected distances called flyrock, are the subject of this work. However, before describing the details, it is good to understand what are the objectives of blasting?

1.2 BLASTING OBJECTIVES

As mentioned earlier, the major objectives of blasting are to break the rockmass, displace it from its in situ position, and to throw the broken rock fragments up to a desired distance and a heap of proper shape, for efficient loading and hauling (Figure 1.2).

The rockmass fragmented by blasting needs to be loaded and transported in an economical way. Hence, the two outcomes of blasting, viz. fragmentation and heave, are of prime importance to a blasting engineer, as these define the economics of a mining operation. This means that the broken material should be of required size and the blasted muck should be casted in a profile that is favourable to the loading equipment, and, in relevant terms, they are in tune with the requirements of the mining subsystem or the system.

Accordingly, blasting cannot be seen in isolation being part of a complete system and its outcome significantly affects the economics of the downstream operations. Blasting is thus a unit operation of a larger system, generally called as mine–mill fragmentation system (MMFS; Figure 1.3), mine to mill system, or drill to mill system. There are two subsystems of "mine" and "mill" in the said system. Metal and non-metal mines generally operate the full MMFS, but coal mines fall within mine

FIGURE 1.2 The processes of breakage and throw due to blasting in a surface mine. (a) Blast bench. (b) Rock breakage in progress (bench being blasting). (c) Throw of the broken rockmass (muck) to a distance. (d) Throw (distance) and heave or final shape of the broken material.

subsystem only. Owing to the economic requirements and environmental constraints, the objectives of blasting can be further described as follows:

1. Breaking of rock by explosives can yield anything, from dust to huge boulders. Fragmentation requirements vary from mineral to mineral. The basic principle is that large equipment is not made to handle large fragments of muck, but to handle large volumes of the blasted material. Too small fragments will involve excessive handling and also reduce the value of ore, and too large fragments will significantly hamper the productivity (Cunningham, 2019). The specifications of loading equipment and size of crusher generally dictate the requirement of fragment size from blasting and hence their optimization. Fragmentation is generally defined in terms of mean fragment size (k_{50}) and the uniformity index (n) assessed with the help of Rosin and Rammler (1933) distribution or Swebrec distribution (Ouchterlony, 2005).
2. Throw is an important requirement for loading as it defines the muck profile or heave and is generally measured in terms of muck profile angle (α). An optimum muck angle (α_{opt}) is different for different loading equipment and depends on the basic operational mechanism of the loading equipment, i.e. whether the loader digs and loads or just scrapes and loads. The looseness of the muck and angle of the muck will define the performance of the loading equipment directly and the hauling equipment indirectly. One of the important aspects of throw of the material during blasting and its impact on ore dilution is described in detail by Gilbride et al. (1995).
3. The presence of people, structures, and other facilities, referred to as objects of concern (OCs) and defined later, within and outside the mine is of concern during blasting. The proximity of such object(s) constraint the blasts in terms of weight of explosives used in a hole and in a delay to control ground vibration and air overpressure within the stipulated limits. Such limits are dependent on excitation frequency of the vibrations and the nature of the structure influenced by blast vibrations. The explosive quantity used in a blasthole also influences the flyrock travel distance.

Moreover, there are conflicts in the cost equations of the unit operations (Calnan, 2015; Comeau, 2019; Mackenzie, 1966; Ouchterlony et al., 2004) and hence a mine–mill fragmentation system (MMFS; Figure 1.3), as defined by Hustrulid (1999a), demands optimization.

There are several works of interest that have provided various methods and means to define the MMFS optimization that in general translates into blast fragmentation optimization. Few such references along with some case studies are compiled in Table 1.1 for the inquisitive reader.

To achieve the fragment size determined by the system, it is imperative to have a proper blast design that yields the economically viable fragment size. It is important to understand that the philosophy of a blast design varies significantly for underground and surface blasting. In the case of underground blasting, a free face must be created, as it is available in surface (mine or bench) blasting.

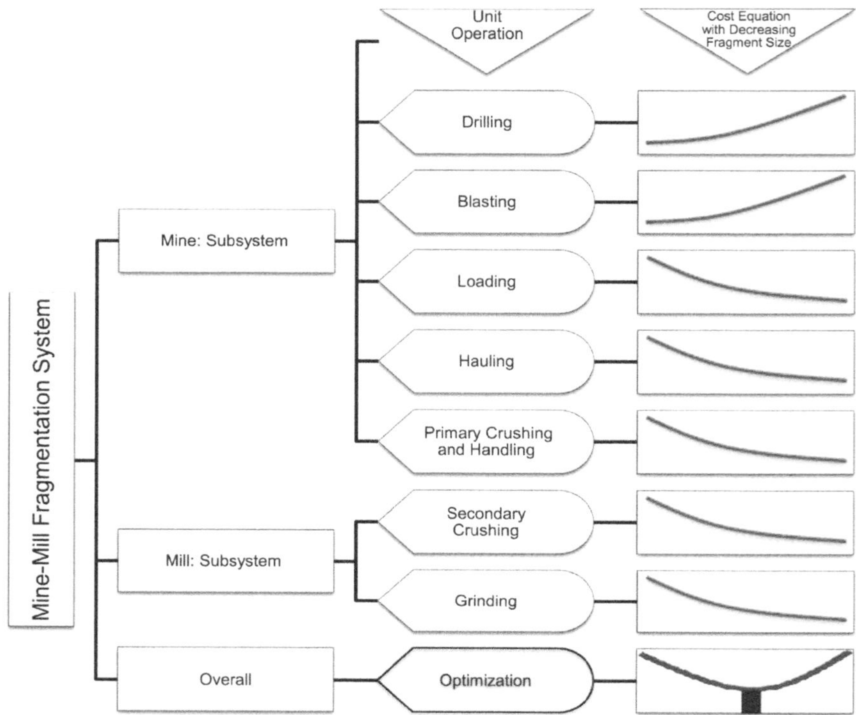

FIGURE 1.3 Mine–mill fragmentation system explained; the alignment of the unit operation arrows points to the change in the cost of the unit operation with change in fragmentation from small to large size (representative trends only).

A significant number of books, texts, and publications, available online and offline, exist on methods of blast design engineering and system optimization. There are quite a few online platforms that claim to work towards optimization of fragmentation through proper engineering and database management systems. Surface blast principles are, however, dealt with in detail by Hustrulid (1999a, 1999b) that include almost all the design considerations, particularly, fragmentation, heave, role explosives, and accessories like delays in the process of rock breakage. On the face of it, one may feel that the design process is intricate and quite complex. However, it will not be out of place to mention that the design process of blasting is not as complicated as it appears. An attempt is hence made here to understand the design process in simple terms.

1.3 BLAST DESIGN

The basic requirements of productivity of a mine are very simple to grasp as the annual production and availability of working hours and equipment can be simply put to work to define the production requirements of a mine, as explained in Figure 1.4.

TABLE 1.1
Some Important Mine–Mill Fragmentation System Optimization Works by Few Researchers

S. No.	Citation	Focus	Details
1	Mackenzie (1966)	Blast optimization	This is a classical study defining and setting the trend for blast fragmentation optimization. The concept was further analysed and detailed by Hustrulid (1999a)
2	Hagan and Just (1974)	Rock breakage theory and optimization	Discusses in detail the breakage by explosive, the role of rockmass, blast design, and explosives on rock fragmentation; muckpile and its impact on economics
3	Morrell and Munro (2000)	JKMRC model application	Describes the use and prediction capabilities of the model for MMFS optimization in three mines
4	Grundstrom et al. (2001)	MMFS optimization	A case study of Porgera gold mine reporting significant improvements in productivity
5	Scott et al. (2002)	MMFS optimization constraints	Reviews the literature on fragmentation optimization. Stress the proper identification of cost-savings in different unit operations in mine-to-mill process and role of experienced manpower to handle such analysis
6	Jankovic and Valery (2002)	Case study on MMFS optimization	Used extensive data to demonstrate the role of analytics and blasting data for cost optimization and report 4–5% increase in the mill by using such strategies
7	Chakraborty et al. (2004, 2005)	Fragmentation evaluation	Comprehensive studies on evaluation of rockmass and blast design variables and their relative importance in defining the fragmentation during blasting and the system optimization routine
8	Esen et al. (2007)	MMFS optimization	Defined a method called process integration and optimization involving benchmarking, rock characterization, measurements, modelling/simulation of blasting and comminution processes, and/or material tracking to achieve best throughput
9	Raina (2013)	Basics of fragmentation optimization	Describes the basics of fragmentation optimization and how it is to be achieved
10	Nageshwaraniyer et al. (2018)	Energy-based method for MMFS optimization	A case study of copper mine economic analysis of unit operations. Spectral imaging was used for tracking, material handling network, and stochastic power consumption in mine-to-mill operations. Economic analysis model was developed for cost-saving

(Continued)

TABLE 1.1 (Continued)

Some Important Mine–Mill Fragmentation System Optimization Works by Few Researchers

S. No.	Citation	Focus	Details
11	Erkayaoglu and Dessureault (2019)	MMFS optimization using data mining and neural networks	Data mining and use of random forest and adaptive boosting algorithm for optimization of mine-to-mill for control and analysis of drilling- and blasting-related variables influencing the productivity
12	Leng et al. (2020)	Oversize and toe formation	Statistical constitutive model developed to evaluate the formation of oversize fragments and toe formation in blasting. Role of satellite holes assessed with the help of numerical method
13	Messaoud et al. (2020)	Oversize production	Microlevel investigations in rock microfabric properties using XRD and microscopic grain identification methods to define the production of oversize fragments in blasting. Statistical methods used for defining optimization of fragmentation due to blasting
14	Assegaff et al. (2020)	Uniform fragmentation	Uses statistical methods to optimize the fragmentation for obtaining uniform fragment size in blasting
15	Park and Kim (2020)	MMFS optimization using MWD technique	A case study of MMFS optimization using monitoring while drilling (MWD) data. Penetration rates were derived from blastholes to work out the intact rock properties and predict the breakage efficiencies influencing comminution energy. Tensile strength and Bond work index correlated with the penetration rate data for crushing and grinding efficiencies
16	Fang et al. (2021)	Fragmentation modelling	Firefly technique has been used for optimization of blast fragmentation and efficacy of the model discussed in comparison to the other artificial intelligence methods used for such operations
17	Zhang and Luukkanen (2021)	Feasibility of MMFS optimization	Discussed the studies that have been successful and unsuccessful in MMFS optimization owing to evaluation of energy efficiency, microcracks in blasting, and redistribution of energy from blasting to milling. The feasibility of MMFS optimization is discussed

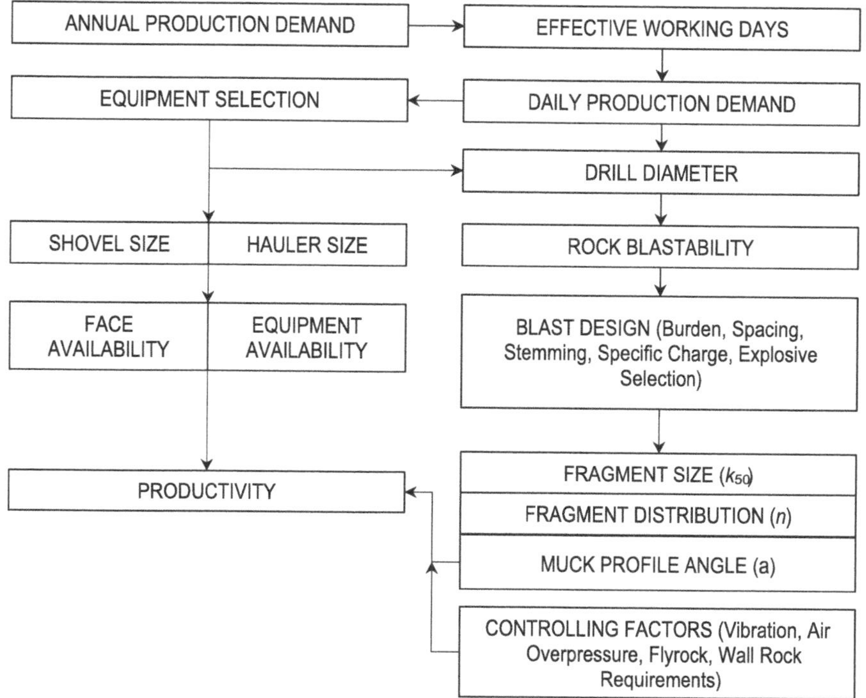

FIGURE 1.4 Basic requirements of productivity in mines. The process involves a comprehensive analysis of requirement of productivity, equipment, and scale of blasting along with its controls.

A brief perusal of Figure 1.4 reveals that the productivity in a mine is pivoted on the blast design, as it determines the degree of fragmentation and hence the productivity of the MMFS. It will not be out of context to explain the terminology, as provided in Figure 1.5, that is important before discussing blast design. As far as possible, the definitions given hereunder, their explanations, and symbols have been adopted as recommended by the ISRM (Rustan et al., 2011).

Free Face

"Free face, an unconstrained surface almost free from stresses, e.g., a rock surface exposed to air or water or buffered rock that provides room for expansion upon fragmentation. Sometimes also called open face."

Blasthole

"Blasthole, a cylindrical opening drilled into rock or other materials for the placement of explosives." Please note that it is used as a single term and not as "blast hole."

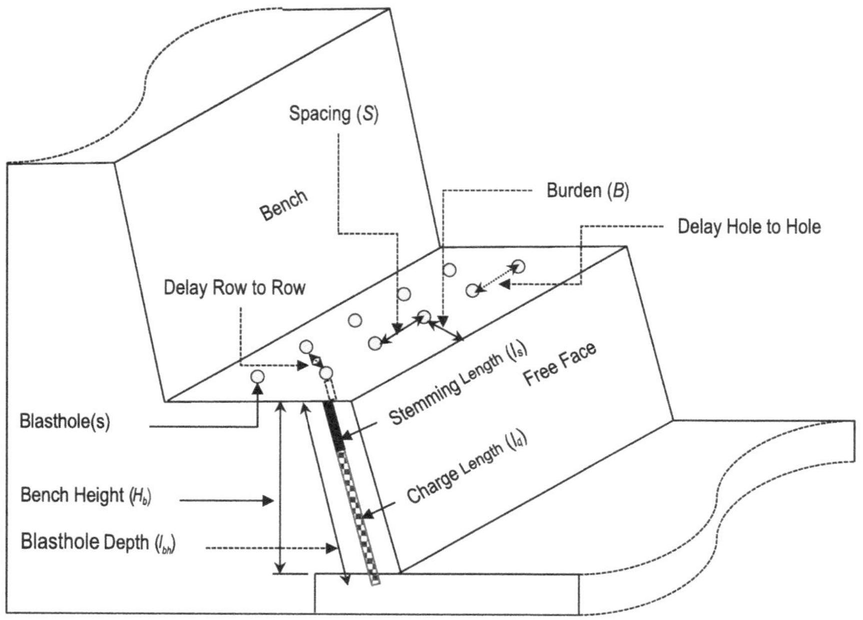

FIGURE 1.5 Basic design variables of a blast in surface mine operations.

Blasthole Inclination or Angle

"Blasthole inclination (α_{bi}), (°), the angle between the blasthole and a reference plane, normally the horizontal plane."

Blasthole Length

"Blasthole length (l_{bh}), (m), the length of the blasthole as measured along the axis from the collar to the bottom of the hole."

Bench Height

"Height of bench (H_b), (m), the vertical distance between the floor and top level of a bench."

Subdrilling Length

"Subdrill, length of blasthole drilled below the planned level of breakage at the floor in bench blasting. subdrilling length (l_{sub}), (m), the length of subdrilling."

Burden

"Burden in bench blasting (B), (m), shortest perpendicular distance between the centre line of a charge and the free or buffered face."

Burden has several variants as follows:

"Burden in crater blasting, blasted burden (true or effective burden) (B_b), (m), critical burden in bench or crater blasting (B_c), (m), drilled burden (B_d), maximum burden (B_{max}), optimum burden (B_{opt}), optimum breakage burden (B_{optb}), optimum fragmentation burden (B_{optf}), practical burden (B_p), reduced burden (B_r).

For explanations, please see Rustan et al. (2011).

Spacing

"Spacing (S), (m), distance between boreholes in a row. It is necessary to distinguish between spacing in drilling (S_d) and spacing in blasting (S_b)." The variants are as follows:

"Spacing in blasting (S_b), (m), the distance between holes initiated on the same delay number. In some blasts when all holes are initiated with different delays the spacing in blasting is defined as the distance between holes detonated consecutively. Spacing in drilling (S_d), (m), the distance between adjacent holes in a row of holes located parallel to the blast front or free surface."

Stemming

Stemming length (l_s) (m), the length of stemming, is the length of the blasthole at its collar that is not filled with explosives and is packed with inert material as defined herein:

Stemming, the inert material of dense consistency, such as drill cuttings, gravel, sand, clay, or water in plastic bags, which is inserted in the collar of the drill hole after charging and used to seal the hole temporarily in order to prevent venting of gas, increase blasting efficiency, to reduce air shock waves or dampen any open flames. In coal mining water stemming cartridges work very well as they also contribute to minimise dust and fires. Stemming is also used as a material to separate explosive charges in a borehole (decks). Stemming can also be used to seal off open cracks intersecting the blasthole.

Specific Charge

Specific charge (q) (kg/m^3) or (kg/t) is the consumption (planned or actual) of explosive per cubic metre or metric ton of rock.

Hole-to-hole Delay

Hole-to-hole delay (t_{HH}) is the time delay used between two adjacent blastholes in order to fire these at different timings.

Row-to-row Delay

Row-to-row delay (t_{RR}) is the time delay used between adjacent rows so as to fire the holes at different timings.

Objects of Concern

Objects of concern (OCs) are objects that are amenable to blast-induced damage, injury, or fatality and are detailed later.

Blast Danger Zone

Blast danger zone (BDZ) is the zone around blasting, generally stipulated or fixed by regulatory authorities, that must be secured and cleared of OCs before blasting. Special permissions are required to blast within BDZ.

1.3.1 Understanding the Blast Design

To design a simple blast, a few rules may be sufficient to understand the same. The process will include the following:

1. Assessing the rockmass blastability preferably in terms of its density and p-wave velocity that define its impedance. There have been attempts to classify rockmass for blasting in terms of rock factor or rock constant "c" by umpteen number of authors. However, a comprehensive review of blastability can be traced to Salmi and Sellers (2021). A scheme for assessing the blastability is also provided in Section 1.3.2.
2. Fixing of the drill diameter which is generally dictated by the bench height and production requirements. However, if a drill with prefixed diameter is already in place, it will constraint the design process.
3. Selection of explosive by knowing its characteristics like its density (ρ_e) and velocity of detonation (c_d). This will be determined by the rock type. A broad criterion is to match the explosive impedance ($c_d \times \rho_e$) with the impedance of the rock, i.e. the product of density of rock and p-wave velocity ($\rho_r \times c_p$).
4. Predicting the initial fragment size through established equations of mean fragment size (k_{50}) and uniformity index (n) of the fragment size distribution.

Thus, the drill diameter (d) in conjunction with the blastability of the rockmass and fragmentation size requirements defines the blast design since one of the major blast design variables, burden (B), is directly related to it (Ash, 1990; Konya & Walter, 1991)—see Figure 1.4.

The variables of blast design that get defined in the process include the burden (B), the spacing (S), and the stemming length (l_s):

$$B = k_b \times d \tag{1.1}$$

TABLE 1.2
Representative Values of Parameters for Major Blast Design Variables

Parameter	Minimum	Average	Maximum	Comments, Use
k_b	25	–	40	Minimum for very hard rock and maximum for highly jointed soft rock
k_s	1.0	1.25	2.0	Minimum for uniform and controlled fragmentation, average for general blasts, and maximum for rockmass that requires simple dislodgement
k_{ls}	0.6	0.7	1.0	Minimum generally not recommended if flyrock is a concern, average for general blasts, and maximum for controlled blasts

$$S = k_s \times B \tag{1.2}$$

$$l_s = k_{ls} \times B \tag{1.3}$$

The practical values and the range of parameters of k_b, k_s, and k_{ls} in Equations 1.1, 1.2, and 1.3, respectively, are given in Table 1.2.

Another simple method for blast design, that incorporates density of rock and explosive, corrections for number of rows, geological conditions, and other conditions, is provided by Konya (1995). Once the design is ready, the fragmentation can be evaluated with the help of Equation 1.4, called the Cun–Kuz model (Cunningham, 2005):

$$k_{50} = c \times q^{0.8} \times Q^{1/6} \times \left(\frac{115}{S_{wr}}\right)^{19/30} \tag{1.4}$$

where k_{50} is the mean fragment size (cm), c is the rock factor or constant varying between 0.2 and 22 and can be estimated with the help of Equation 1.8, q is the specific charge (kg/m^3) defined as the ratio of total explosive used in kg to the total volume of blast in terms of product of B, S, and bench height, Q is the explosive weight (kg), and S_{wr} is the relative weight strength of explosive relative to ANFO (ammonium nitrate fuel oil).

With moderate changes in blast design and experimentation thereof, the optimum fragment size and blast design can be identified to draw a production pattern. Simulations or trial and error method with proper planning can be put to work for establishing the best pattern along with blast design iterations (Hustrulid, 1999a). Further requirements of throw and reduction of other unwanted effects can be taken into consideration by changing the hole-to-hole and row-to-row delays and even modifying the blast design. The aforementioned method will require measurement of fragmentation that is economically viable.

1.3.2 Rockmass and Blast Design

Rockmass is a very complex subject as it involves several inherent properties that puzzle the excavation engineer. Engineers, however, need to predict the outcome of the blast process that demands conversion of the rockmass and explosive into numbers. However, explosive characteristics can be achieved by measurements, as explained further in Section 1.3.3. Putting rockmass to numbers is not only a very subjective matter but also a complex topic. The inherent variability of rockmass in terms of its strength, jointing, and microscopic properties makes it fuzzy and qualitative in nature. Frequent spatial changes of rockmass properties add to the difficulty of rockmass characterization. Consequently, a host of classifications, methods, and rating systems for rockmass blastability, that defines the ease with which a rockmass can be fragmented by blasting, have been developed over the years. To have a broad idea of the complexity in rockmass classifications and blastability, a list of rockmass properties (both macroscopic and microscopic), their nature, and the feasibility of converting them to numbers is presented in Table 1.3.

The role of microscopic properties in blasting and their incorporation in classification systems is practically lacking or has been confined to laboratory or numerical analytical studies only determining the strength of the intact rocks (Abdlmutalib & Abdullatif, 2019; Ahmad et al., 2017; Dobereiner & de Freitas, 1986; Jeng et al., 2004; Kamani & Ajalloeian, 2019; Messaoud et al., 2020; Tsidzi, 1990).

The macroscopic properties of rockmass have been treated fairly that resulted in summation of such individual properties with development of rating or classification systems. Accordingly, there have been multiple attempts to group these into classification systems to define the classes of the rockmass for ascertaining stability of the rockmass in underground excavations by employing the macroscopic variables. Some such classifications that have been widely used for mining and excavation purpose are rockmass rating or RMR (Bieniawski, 1989), rock quality designation (RQD) (Deere & Deere, 1989), index of rock quality or Q (Barton et al., 1974), mining rockmass rating (MRMR) (Laubscher & Jakubec, 2001), geological strength index (GSI) (Marinos et al., 2005), and rockmass index or RMi (Palmström, 1996). These classification systems are functions of strength of intact rock and adjusted for block volume or joint density, joint roughness, joint alteration, joint size, and many other such properties. RMi is probably the first such classification that incorporates blastability of rockmass but has not been used for the purpose because of complexity in calculations. Aforementioned systems are known as geomechanics rockmass classifications.

There are some examples of use of rockmass classifications for defining the outcome of blasting, but such studies are restricted to quality of blasting (Innaurato et al., 1998) impact on controlled blasting (Singh, 2003), prediction of overbreak (Jang & Topal, 2013; Koopialipoor et al., 2019; Segaetsho & Zvarivadza, 2019), and blast overpressure (Gao et al., 2020) in mining or civil excavations. Also, such studies are negligible and have not been widely accepted by field engineers for defining the blastability of the rockmass. Adhikari et al. (1999) based on a significant database of an underground cavern contested the findings of Ibarra et al. (1996) that the performance of underground blasts like overbreak and underbreak can be compared

TABLE 1.3
Comprehensive Evaluation of Rockmass for Engineering Purpose and Problems Faced in Quantification of Rockmass

Macroscopic Properties	Nature/Comments	Microscopic Properties	Nature/ Comments
Intact rock strength	Quantifiable, compressive, and tensile strengths of intact rock can be determined in laboratory	Grain size	Quantifiable
Rockmass strength	Difficult to quantify owing to the presence of joints and joint properties	Grain fabric	Fuzzy
Lithology	Varies significantly over space, presents difficulties in characterization in linear excavations like tunnels	Grain boundary	Fuzzy, fractal
Number of joints	Quantifiable	Boundary strength	Difficult to quantify
Joint spacing	Quantifiable, multiple joints association create confusion. Which joint to be considered is an issue	Grain shape	Quantifiable but difficult to include in definitions
Weathering	Qualitative, quantification is not practical		
Joint length	Quantifiable		
Joint strength	Quantifiable, cohesion, and friction angle needs detailed testing		
Joint alteration	Qualitative, difficult to put to numbers		
Joint filling	Qualitative, difficult to put to numbers		
Joint aperture	Quantifiable, fuzzy		
Joint roughness	Fuzzy, fractal		
Block size	Quantifiable, varies over a wide range, restrictions due to joints occurring in different planes		
p- and s-wave velocities	Quantifiable, have significant role as blasting, and these properties are of dynamic nature. Define impedance, in situ determination requires a lot of expertise		

Fuzzy indicates the variations are not crisp but overlap in different classes
Fractal means that the distributions have fractal nature

to rockmass quality "Q" of Barton et al. (1974). Only a few works report the use of such classifications in flyrock modelling, e.g. use of *RMR* (Bakhtavar et al., 2017; Hasanipanah & Amnieh, 2020; Monjezi et al., 2011, 2012; Wu et al., 2019) and GSI (Asl et al., 2018).

Although some of the properties considered in such classifications can be used in assessing blastability, the use of such classifications in totality needs further evaluation, despite the fact that some recent studies (Nur Lyana et al., 2016; Sayevand & Arab, 2019) have used such classifications for defining fragmentation. In order to compare the role of rockmass on fragmentation, Doucet et al. (1996) conducted limited number of experiments. Despite no significant relationship, they observed that the characteristic particle size after blasting (k_c blast) increases when the characteristic size of the in situ distribution (k_c in situ) increases and that k_c of blast decreases when the adjusted powder factor or specific charge increases.

The energy-block-transition model (Lu & Latham, 1998) defines blastability through a comprehensive mathematical treatment of transformation of in situ block size to fragmented block size. They considered 12 factors, viz. uniaxial compressive strength, uniaxial tensile strength, density of rock, static or dynamic modulus of rock, p-wave velocity, Schmidt hammer rebound value, Poisson's ratio, fracture toughness of rock, mean in situ block size, fractal dimension of in situ block sizes, wave velocity ratio, the ratio of p-wave velocity in field to that in laboratory or by rock quality designation (RQD), and cohesion or friction angle of discontinuity plane for defining the blastability. However, they (Lu & Latham, 1998) ignored the role of joint orientation, although Chakraborty et al. (1994) had observed that the mean fragment size along with depth and cross-sectional area of broken zone were significantly impacted by the joint orientation.

Zhang (1990) developed a fivefold classification system for blastability based on structural types of rockmass, characteristics of crustal stress, and blasting vibration effect. Kiliç et al. (2009) proposed a model for blastability in terms of tensile strength and coefficient of internal friction. Tsiambaos and Saroglou (2010) proposed a classification method for rockmass excavatability based on GSI but they did not present any formula for the calculations. Split Hopkinson pressure bar tests on artificial joints in blocks were conducted by Li et al. (2016) who observed that the deformation of rockmass is caused by the joint deformation and the closure volume of joints increases when contact area declines. Choudhary et al. (2016) reported the effect of rockmass properties on blast fragmentation and concluded that with increase in porosity, compressive strength, and size of the in situ blocks, the fragment size decreases but increases, if the density of rock increases.

Also, the rockmass, blast design, and explosive properties vary in a mine and results in variation in blast outcomes like fragmentation. If such variables are treated as distributions, the resultants conform to the measured results (Thornton et al., 2002). The method proposed by Thornton et al. (2002) provides a basis for revisiting the existing method of reporting and analysis of blast input, output variables, and related analysis. McKenzie et al. (1982) attempted to quantify rockmass variables for modelling of fragmentation using a cross-hole acoustic method incorporating propagation velocities of waves and their attenuation. Nevertheless, Sellers et al. (2019) concluded that there is no commonly accepted method of rock blastability which should include rockmass strength, fracture frequency, and density. They supported the use of seismic velocities for such classifications as these present a holistic picture of the rockmass.

However, the classification system that defines blastability specifically while using the macroscopic properties of the rockmass was introduced by Lilly (1986). This classification, popular despite its shortcomings, uses a rating for different factors used to define a blastability index (I_{BI}) and is given in Table 1.4.

TABLE 1.4
Ratings Used in Lilly's (1986) Classification System for Defining Blastability Index

Description	Rating	Details
RMD (rockmass description)	10	Powdery/Friable rockmass
	20	Blocky rockmass
	50	Totally massive rockmass
JPS (joint plan spacing)	10	Close spacing (<0.1 m)
	20	Intermediate (0.1–1.0 m)
	50	Wide spacing (>1.0 m)
JPO (joint plane orientation)	10	Horizontal joints
	20	Dip out of the face
	30	Strike normal to face
	40	Dip into face
SGI = Specific gravity influence	25 × SG − 50	SG is the specific gravity of rock (t/m^3)
H = Hardness	1–10	Mohs scale

The blastability index (I_{BI}) can be calculated by substituting the ratings in Table 1.4 in Equation 1.5:

$$I_{BI} = 0.5 \times \left(RMD + JP + JPO + SGI + H \right) \tag{1.5}$$

The specific charge (q) and the energy factor (E_f) can be worked out from Equations 1.6 and 1.7, respectively.

$$q = 0.004 \times I_{BI} \tag{1.6}$$

$$E_f = 0.015 \times I_{BI} \tag{1.7}$$

The rock constant or factor mentioned in Equation 1.4 can be estimated by modifications in Equation 1.5 and is given in Equation 1.8.

$$I_{BI} = 0.06 \times (RMD + SGI + H) \tag{1.8}$$

Bameri et al. (2021) presented a case study of application of I_{BI} (Lilly) in a copper mine while using Monte Carlo simulation and found that the method provided a better insight into the combinatorics of rockmass factors. The use of I_{BI} in the prediction of fragmentation and wall control is documented in Chung (2001), Chung and Katsabanis (2000), Monjezi et al. (2011), Segaetsho and Zvarivadza (2019), and many other such publications.

TABLE 1.5
Blastability Classification of Rockmass Incorporating Adjustments for Confinement and Stiffness (Ghose, 1988)

Variable	Range				
Density, t/m^3	<1.6	1.6–2.0	2.0–2.3	2.3–2.5	>2.5
Rating	20	15	12	6	4
Spacing of discontinuity, m	<0.2	0.2–0.4	0.4–0.6	0.6–2.0	>2
Rating	35	25	20	12	8
Point load strength index, MPa	<1	1–2	2–4	4–6	>6
Rating	25	20	15	8	5
Joint plane orientation	Dip into face	Strike normal to face	Horizontal joints	Dip out of face	Strike at an acute angle to face
Rating	20	15	12	10	6
Adjustment factor, A_{F1}	Highly confined, Rating = −5 Reasonably free, Rating = 0				
Adjustment factor, A_{F2}	Hole depth to burden ratio >2 0, Rating = 0 Hole depth to burden ratio = 1.5–2.0, Rating = −2 Hole depth to burden ratio <1.5, Rating = −5				
Blastability Index	70–85	60–70	50–60	40–50	30–40
Specific charge (kg/m^3)	0.2–0.3	0.3–0.5	0.5–0.6	0.6–0.7	0.7–0.8

Lilly's (1986) classification system was further modified by Ghose (1988) by incorporating influence of confinement and stiffness that is defined in terms of hole depth to burden ratio (Equation 1.9):

$$I_{BI} = \rho_r + J_s + I_{PL} + J_{PO} + A_{F1} + A_{F2} \tag{1.9}$$

where ρ_r is the density of rock in t/m^3, J_s is the joint spacing in m, I_{PL} is the point load strength index, J_{PO} is the joint plane orientation, and A_{F1}, A_{F2} are adjustment factors (Table 1.5). The blastability index values and corresponding specific charge can also be seen in Table 1.5.

The classification of Ghose (1988) is an improvisation of Lilly's (1986) original method as it incorporates some design aspects relating to hole depth, burden, and confinement of a blast. Other blastability assessment methods have been compiled in Table 1.6.

Salmi and Sellers (2021) summarized most of the developments in blastability through a comprehensive review that found the dynamic breakage of rockmass influenced by the strength, density, and structure of the discontinuities in the rockmass. They held that in situ block size defines the fragmentation, attenuation of stress waves, and the extent of damage zone about blastholes. They identified a comprehensive list

TABLE 1.6
Few Other Blastability Assessment Methods

S. No.	Citation	Equation	Description
1	Muftuoglu et al. (1991)	$q = 0.0025\sigma_t^2 - 0.0042\sigma_t + 0.1363$	Specific charge (q) in kg/m³ is defined as second-order equation of tensile strength (σ_t)
2	Scott (2020)	$f_d = 0.0371 \times \rho_r^2 - 0.0512\rho_r + 0.9172$	Density factor (f_d) which is a quadratic function of density (ρ_r) is in T/m³
3	Borquez (1981)	$c = 1.96 + 0.27 \times \ln(ERQD)$	Blastability factor (c), equivalent rock quality designation (ERQD) that considers joint strength and alteration
4	Da Gama (1995)	$C = 15424c^2 - 2840.6c + 146.27$	Blastability (c), cohesion (C) is in MPa
5	Scott and Onederra (2015)	$f_s = 0.0549 \times \sigma_c^{0.5315}$	Strength factor (f_s) as a function of compressive strength (σ_c) is in MPa
6	Rakishev (1981)	$\sigma_s = 0.1\sigma_c + \sigma_t$	Limit strength of the rocks (σ_s) in MPa is expressed in terms of compressive and tensile strengths
7	Kou and Rustan (1992)	$c = \dfrac{\sigma_c^2}{2E_r \times \eta \times Q_e}$	Blastability factor (c) is given in terms of σ_c, detonation heat (Q_e) in kJ/kg, Young's modulus of rock (E_r) in MPa, and energy transformation efficiency (η)
8	Zou (2016)	$N = 67.22 - 38.44Ln(V_c) + 2.03Ln(\rho_r C_p) + K$	Rock blastability is defined in terms of volume of crater (V_c) in m³, p-waves velocity (c_p) in m/s, ρr in ×103 kg/m³, index of rock fragmentation (K)
9	Qu et al. (2002)	$f_e = a_1\rho_e c_d^2$ $c = a_2(\rho_r\sigma_c^\alpha)\log(a_3 S_{javg})^\beta$	Explosive strength factor (f_e), ρ_e in ×103 kg/m³, constants which can be determined from regression analysis ($a1$, $a2$, $a3$, α, β), velocity of detonation (c_d) in m/s, blastability (c), average joint spacing (S_{javg}) in $\times 10^{-2}$ m, σ_c in $\times 10^{-3}$ MPa
10	Livingston (1956)	$B_{opt} = k\sqrt[3]{Q}$	Optimum breakage burden distance or charge depth (B_{ob}) in m, k is a constant of proportionality expressing rock and explosive properties, Q is the mass of explosive in kg
11	Dick et al. (1990)	$\log\sqrt{(B_{opt}^2 + r_a^2)} = 1.846 + 0.312\log Q$	Optimum breakage burden distance is a function of apparent crater radius (r_a) and the equivalent TNT charge mass (Q)

of almost 32 blastability assessment methods found in the literature and ranked these for establishing a three-dimensional classification system for blastability. However, the universal acceptance of any of the methods, defining blastability, is still lacking.

1.3.3 Explosive and Blast Design

Explosives properties play a major role in defining the outcome of the blast (Božić, 1998). A comprehensive detail of explosives, their use, and some basics related to the outcome of blasts are provided by Cooper (1996). Commercial explosives are mixtures and hence cannot be directly tested efficiently for their energy yield. Fragmentation efficiency is based on ratings that are given by explosive manufacturers through proprietary codes (Djordjevic, 2001). This limits the comparison of results of explosives of different manufacturers. Djordjevic (2001) developed a code named CHEETAH for explosive standardization and concluded that the code can be used further for explosive selection on optimal basis. Owing to the complexity in such calculations of energy, users generally resort to simple assessment of velocity of detonation and density of explosives that yields its impedance and have been correlated for determination of blasthole pressures (Cooper, 1996) as provided in Equation 1.10:

$$P_b = \frac{\rho_e \times c_d^2}{8} \tag{1.10}$$

where P_b is the blasthole pressure, $\times 10^3$ MPa, and is generally considered to be 0.5 times the detonation pressure; ρ_e is the density of explosive, kg/m^3; and c_d is the velocity of detonation of explosive, m/s.

As mentioned earlier, the best way of selection of explosive is to match the impedance of rock with that of the explosive. There is a further scope of improvement in Equation 1.10 since density and velocity of detonation of explosive are related significantly (Cunningham, 2006).

Trials in blocks of granite, porous limestone, and sandstone with different explosives were conducted by Bergmann et al. (1973) and they correlated energetics of explosive with average fragment size, burden velocities, and peak pressures. Bergmann et al. (1973) derived that the fragmentation is controlled by explosive energy, its detonation velocity and density, and the degree of coupling, a ratio of explosive diameter to blasthole diameter. Although the study was conducted on blocks and may not reflect the impact of joints, the study provides a basis for determination of such relationships and need to be extended to full-scale blasting in actual bench conditions. However, such conclusions were contested by Cunningham (2006) citing that the results were due to the size of blocks and not due to the velocity of detonation. Later, Agrawal and Mishra (2017) found that the velocity of detonation of an explosive, under specific rock and test conditions, can be used to evaluate the performance of the explosive used in mines.

A method to determine the pressure fluctuations adjacent to the blasthole due to gas was developed by Williamson and Armstrong (1986). The authors claim that the method can be used for explosive selection and for optimization of directional blasting and control of damage to the parent rock.

Three different explosives were used to test fragmentation results, under controlled conditions, in granite lumps to model the post-blast residual rock fragment strength that were considered to improve the SAG mill productivity by 20% (Michaux & Djordjevic, 2005). Ning et al. (2011) attempted to model rockmass failure mechanism and gas penetration effects using discontinuous deformation analysis for artificial joints. Mishra and Sinha (2016) conducted 20 trail blasts with emulsion while varying explosive density from 1000 kg/m^3 to 1170 kg/m^3 and concluded that the explosive with higher density yielded better fragmentation with average fragment size reduction ratio of 1.45 as compared to the low-density explosive. Mishra and Sinha (2016) estimated 27% cost-saving in overall cost of the mining process in the study with implementation of high-density explosive. Inanloo et al. (2018) developed a model for the prediction of k_{50} and observed that specific energy of explosive and burden are the most important variables in determining the fragmentation.

Khademian and Bagherpour (2017) conducted tests with three types of explosives to estimate the effect of explosive type on post-blast strength of rock fragments, their microcracks content, and energy utilization in successive processes. They reported that with the use of pentolite, the rock strength reduced by 16.5% and that the strength of blasted fragments highly correlates with velocity of detonation of explosives and increases the microcrack density greater than 300%. Mertuszka et al. (2018) also concluded that blasthole diameter has a significant influence on continuous velocity of detonation of explosives.

Despite the significant work on commercial explosive selection and performance, there is a disagreement in authors about the best measure of explosive property that could be used for determining the blast outcome. It is expected that in absence of such a factor, the velocity of detonation and density of explosives will continue to dominate the discussion space.

1.4 BLAST OUTCOME: THE NEGATIVES

The outcome of a blast may be divided into two groups, viz. desired and undesired, as explained in Figure 1.6. The desired outcomes like fragmentation, throw, and heave have been explained earlier. The undesired outcomes of blasting include ground vibrations, air overpressure and noise, noxious fumes, backbreak and side-break, and dust and flyrock that need attention of a blasting engineer. These unwanted resultants of blasting cannot be done away with as they are complementary to the process but can be minimized. This work is focused on one such prevalent negatives of blasting, i.e. flyrock, and to understand several dimensions of the phenomenon, its generation, reasons for occurrence, its probabilities along with associated risks, and control measures. Flyrock is a rock fragment of any size emanating from a blast, travelling few tens to hundreds of metres from the blast with potential threat to the property or people, as is further detailed in Chapter 2. Unfortunately, flyrock has received little attention from researchers in comparison to ground vibrations for different reasons (Raina, 2014) and is still not the favourite subject of researchers.

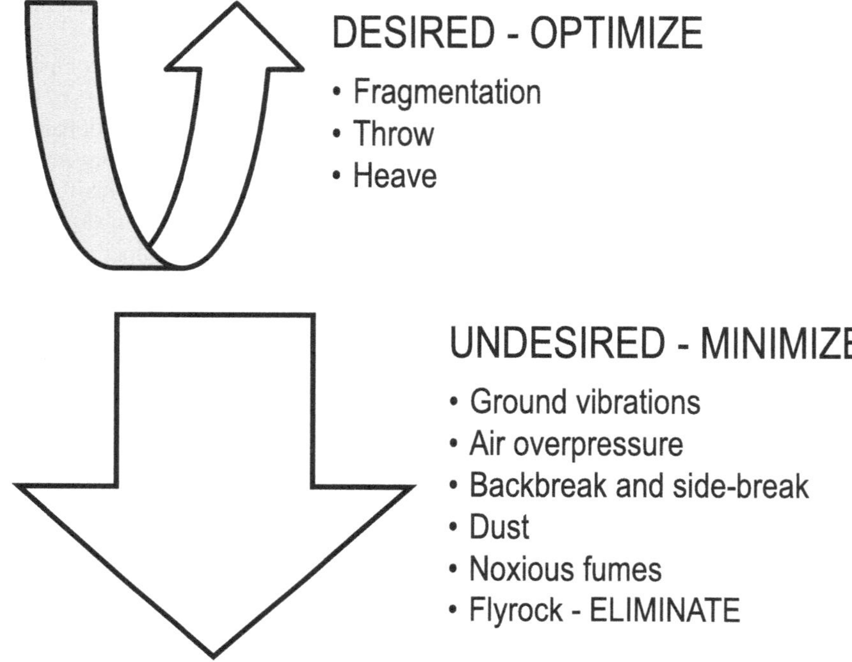

FIGURE 1.6 Main outcomes of blasting—objectives are to optimize, minimize, or to eliminate.

1.5 FOCUS FLYROCK: WHY?

Despite the advancement in understanding of the blasting in surface mines, flyrock has intrigued the mining fraternity and the blasters day in and day out, for quite some time. There were times when people did not bother about the occurrence of flyrock for the following reasons:

1. The mines were quite far away from the blast zones, i.e. in current scenario, the mines have expanded spatially and/or human settlements have transgressed the mining space domain.
2. Other issues like the presence of wildlife and their proximity to mining activity were not a concern earlier, but have also taken a centre stage.
3. The laws governing the blasts were not so stringent. However, in recent times, things have changed remarkably.

The situation in recent times has however changed primarily due to the following facts:

1. Flyrock is not just associated with mining only.
2. With the growing demands of urbanization, infrastructure development, and hydroelectric projects in construction, flyrock has become a universal phenomenon.

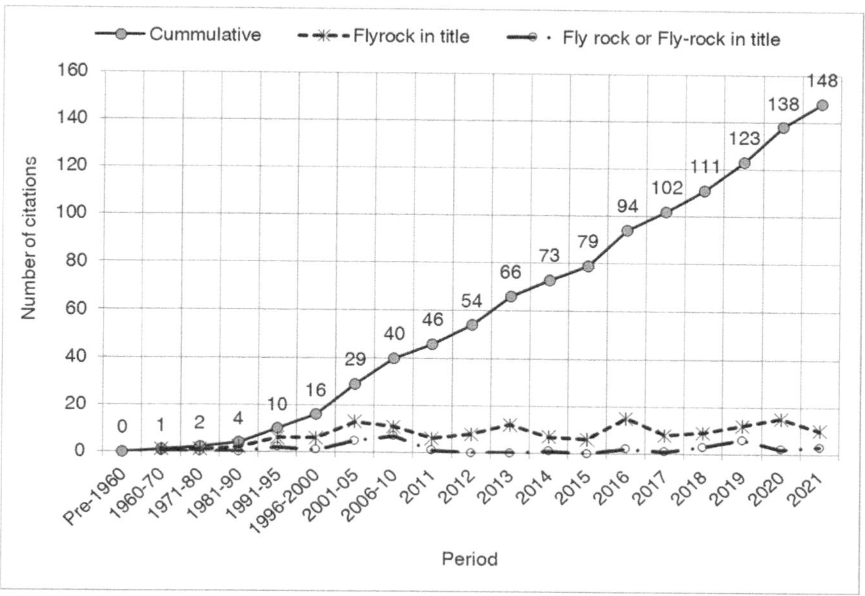

FIGURE 1.7 Number of publications with "flyrock," "fly rock," or "fly-rock" in the title in different periods.

Source: Google Scholar.

3. The definitions of "blast danger zone" (BDZ, the zone around a blast that must be cleared of OCs before blasting) have changed and vary from country to country.
4. Any flyrock incident can even completely halt the mining or any construction project operations.

The fact remains that the issue of flyrock has been out there, but reporting of such incidents has been minimal. In practice, only flyrock(s) resulting in fatalities are reported. The remaining data on flyrock remains elusive. The science has also advanced much, and in contrast to the earlier view that flyrock occurrence or its travel distance cannot be predicted, the efforts to investigate it in detail have intensified with time, as is evident from Figure 1.7.

Figure 1.7 has been composed from the data acquired through search with the keyword "flyrock," "fly rock," or "fly-rock" in the title from Google Scholar and indicates that there has been an increased effort in investigations despite the following reasons in reporting or investigations (Raina, 2014):

1. The flyrock incidents are underreported because of the possible penalties (Davies, 1995). There are severe consequences of flyrock casted beyond the BDZ defined by the regulatory bodies across the world. Severe restrictions and even closure of mines can be traced to literature or public domain.

2. The phenomenon is distinctly irregular in nature that restricts data generation. One cannot generate flyrock in testing as it could be dangerous. This probably constraints the experimentation in flyrock.
3. Not much data is available on the subject probably due to the previous two reasons. The data reported in a host of latest publications belong to a few mines, lacking probability analysis and other important aspects like size, velocity, and launch angle of flyrock along with minimal mention of the characteristics of the rockmass blasted. Also, there is least collaboration between international agencies to pool the data and deduce tangible results for flyrock prediction and control.
4. There are numerous variables in blasting that may belong to controllable or non-controllable domain (Hustrulid, 1999a) and hence restrict experimentation. Since most blast design variables are dependent on drill diameter, as explained in Figure 1.4, it is difficult to have a controlled experiment. Even if this is made possible, the rockmass property is not possible to control, being uncontrollable in nature. Also, it is very difficult to assign crisp values to blastability of rockmass as the ambiguity in its characterization is yet to be resolved. This is one of the reasons that formulae for the outcome of blasting, rather than pure mathematical equations that are dimensionally balanced, are in vogue (Kou & Rustan, 1992).
5. There are several confusions in understanding of the flyrock phenomenon, particularly the development of relationships of flyrock, just with the use of basic blast design variables. This has created a lot of misunderstanding about flyrock. In addition, there are confusions in the use of flyrock and associated terminology. A lot of citations on flyrock report use of intelligent methods for flyrock predictions despite very weak conceptual foundations (see Section 4.6).
6. Any size of the flyrock can be ejected from the blast face, but most of the publications are silent on the size of flyrock. Also, there is confusion as what size of rock fragment ejected from a blast can be categorized as flyrock. Fine fragments do not travel large distances owing to less surface area and hence have less impact. Most of these are not even noticed. Therefore, it is prudent that flyrock requires to be categorized on logical foundations.
7. The safe distance defined in terms of BDZ is yet a fuzzy boundary. Event-based distances are vague and unscientific and "risk-based" definition specific to a mine adopted uniformly presents a better solution. The question arises if there is a simple formula defining the ground vibration attenuation constants for a mine, why such a criterion is missing for flyrock. Scientists and policy planners should be aware that flyrock is more dangerous than ground vibrations.

Another major observation is that there have been enormous studies on ground vibration due to blasting. These range from simple reports with prediction methods and those with intelligent techniques for control of ground vibrations. A brief survey of such publications is presented in Table 1.7.

TABLE 1.7
Citations on Ground Vibration in Literature that Can Be Traced in Google Scholar or Other Members-only Sites

Keywords in Title	Number of Citations		
	Google Scholar	ISEE (OneMine)	ISEE Only
Blast vibration	632	1029	639
Blasting vibration	978		

Source: ISEE (2022)

The total publication count of blast-induced ground vibration monitoring and control is greater than 2000 and confirms that the research on flyrock is not even 10% of that of the ground vibrations.

One of the reasons for such vast studies on ground vibrations and air overpressure is that these blast outcomes are complementary to blasting alongside the fragmentation, throw, and toxic fumes. This provides an ideal situation for monitoring. You can generate several events of ground vibration from a single blast with the use of multiple seismographs and create a database of such events through limited experiments. Such seismographs have been developed and are still proliferating with advancement of IoT. Also, there are specific guidelines for manufacturing, deployment, and calibration[2] of such seismographs that validate their use, alongside the business interests of the manufacturers. The case with flyrock is, however, altogether different, as explained.

This situation prevails despite the fact the flyrock can be fatal and ground vibrations do not. Despite the aforementioned facts, researchers shy away from investigation flyrock, either for the fact that they have probably understood flyrock, or they are constrained by the conditions of its reporting, testing, and occurrence, as mentioned earlier. Hence, the case for flyrock investigations is justified. The said inferences also point to the fact that the flyrock problem needs to be revisited. We will dive deeper in the subsequent chapters to have a cognizable info base on flyrock. This will mean revisiting the existing predictive regimes to include uncertainties and unknowns in rockmass and possibilities of design mistakes that will do justice to the subject. An international collaboration for data sharing is desirable to devise a simple method to know whether flyrock will occur or not.

NOTE

www.globenewswire.com/en/news-release/2022/04/01/2414830/0/en/Global-Explosive-Market-Size-2022-2028-Industry-On-Going-Trends-Analysis-By-Country-Level-Data-Production-Capacity-Estimates-Economic-Change-Development-History-Regional-Overview-S.html (10.05.2022).

REFERENCES

Abdlmutalib, A., & Abdullatif, O. (2019, March 26–28). Control of pore types and fracture intensity on the P-wave velocity of carbonate rocks. Proceedings of the *International petroleum technology conference 2019* , ITPC, 6p. https://doi.org/10.2523/iptc-19527-ms

Adhikari, G. R., Rajan Babu, A., Balachander, R., & Gupta, R. N. (1999). On the application of rock mass quality for blasting in large underground chambers. *Tunnelling and Underground Space Technology*, *14*(3), 367–375. https://doi.org/10.1016/S0886-7798(99)00052-8

Agrawal, H., & Mishra, A. K. (2017). A study on influence of density and viscosity of emulsion explosive on its detonation velocity. *Modelling, Measurement and Control C*, *78*(3), 316–336. https://doi.org/10.18280/mmc_c.780305

Ahmad, M., Ansari, M. K., Singh, R., Sharma, L. K., & Singh, T. N. (2017). Estimation of petrographic factors of Deccan basalt using petro-physico-mechanical properties. *Bulletin of Engineering Geology and the Environment*, *76*(4). https://doi.org/10.1007/s10064-017-1061-0

Ash, R. L. (1973). *The influence of geological discontinuities on rock blasting*. University of Minnesota.

Ash, R. L. (1990). Designing of blasting rounds. *Surface mining, Society for Mining and Exploration Inc*. (pp. 565–578). Port City Press, Maryland.

Asl, P. F., Monjezi, M., Hamidi, J. K., & Armaghani, D. J. (2018). Optimization of flyrock and rock fragmentation in the Tajareh limestone mine using metaheuristics method of firefly algorithm. *Engineering with Computers*, *34*(2), 241–251. https://doi.org/10.1007/s00366-017-0535-9

Assegaff, F. R., Endayana, C., Sulaksana, N., & Waloeyo, D. P. (2020). Optimization of limestone production based on geological structure conditions for blasting geometry design. *Jurnal Geologi Dan Sumberdaya Mineral*, *21*(2), 69. https://doi.org/10.33332/jgsm.geologi.v21i2.483

Bakhtavar, E., Nourizadeh, H., & Sahebi, A. A. (2017). Toward predicting blast-induced flyrock: A hybrid dimensional analysis fuzzy inference system. *International Journal of Environmental Science and Technology*, *14*(4), 717–728. https://doi.org/10.1007/s13762-016-1192-z

Bameri, A., Seifabad, M. C., & Hoseinie, S. H. (2021). Uncertainty consideration in rock mass blastability assessment in open pit mines using monte carlo simulation. *Eurasian Mining*, *35*(1), 34–38. https://doi.org/10.17580/em.2021.01.07

Barton, N., Lien, R., & Lunde, J. (1974). Engineering classification of rock masses for the design of tunnel support. *Rock Mechanics Felsmechanik Mécanique Des Roches*, *6*(4), 189–236. https://doi.org/10.1007/BF01239496

Bergmann, O. R., Riggle, J. W., & Wu, F. C. (1973). Model rock blasting-effect of explosives properties and other variables on blasting results. *International Journal of Rock Mechanics and Mining Sciences*, *10*(6), 585–612. https://doi.org/10.1016/0148-9062(73)90007-7

Berta, G. (1990). Explosives: An engineering tool. In M. Italesplosivi (Ed.), *Italesplosive* (Vol. 6). https://adams.marmot.org/Record/.b16176315

Bhandari, S. (1997). *Engineering rock blasting operations*. Balkema.

Bieniawski, Z. T. (1989). Engineering rock mass classifications: A complete manual for engineers and geologists in mining, civil, and petroleum engineering. In *Engineering rock mass classifications: A complete manual for engineers and geologists in mining, civil, and petroleum engineering*. John Wiley & Sons.

Borquez, G. V. (1981). Estimating drilling and blasting costs - an analysis and prediction model. *Engineering and Mining Journal*, *182*(1), 83–89. https://doi.org/10.1016/0148-9062(81)90231-x

Božić, B. (1998). Control of fragmentation by blasting. *Rudarsko Geolosko Naftni Zbornik, 10*, 49–57.

Calnan, J. T. (2015). Determination of explosive energy partition values in rock blasting through small-scale testing. *ProQuest Dissertations and Theses, 168*. https://search.proquest.com/docview/1764221237?accountid=188395

Chakraborty, A. K., Jethwa, J. L., & Paithankar, A. G. (1994). Assessing the effects of joint orientation and rock mass quality on fragmentation and overbreak in tunnel blasting. *Tunnelling and Underground Space Technology Incorporating Trenchless, 9*(4), 471–482. https://doi.org/10.1016/0886-7798(94)90106-6

Chakraborty, A. K., Raina, A. K., Ramulu, M., Choudhury, P. B., Haldar, A., Sahoo, P., & Bandopadhyay, C. (2004). Development of rational models for tunnel blast prediction based on a parametric study. *Geotechnical and Geological Engineering, 22*(4), 477–496. https://doi.org/10.1023/B:GEGE.0000047042.90200.a8

Chakraborty, A. K., Ramulu, M., Raina, A. K., Choudhury, P. B., Haldar, A., Sahoo, P., & Bandopadhyay, C. (2005). Rock fragmentation system analysis applying cause-effect relationship. *Journal of Mines, Metals and Fuels, 3–4*, 47–54.

Choudhary, B. S., Sonu, K., Kishore, K., & Anwar, S. (2016). Effect of rock mass properties on blast-induced rock fragmentation. *International Journal of Mining and Mineral Engineering, 7*(2), 89–101. https://doi.org/10.1504/IJMME.2016.076489

Chung, S. H. (2001). An integrated approach for estimation of fragmentation. *Proceedings of the Annual Conference on Explosives and Blasting Technique, 1*, 247–256.

Chung, S. H., & Katsabanis, P. D. (2000). Fragmentation prediction using improved engineering formulae. *Fragblast, 4*(3–4), 198–207. https://doi.org/10.1076/frag.4.3.198.7392

Comeau, W. (2019). Explosive energy partitioning and fragment size measurement—importance of the correct evaluation of fines in blasted rock. In B. Mohanty (Ed.), *Proceeding of the workshop on measurement of blast fragmentation, Fragblast-5* (pp. 237–240). Balkema. https://doi.org/10.1201/9780203747919-33

Cooper, P. (1996). *Explosives engineering*. John Wiley & Sons.

Cunningham, C. V. B. (2005). The Kuz-Ram fragmentation model—20 years on. In *Brighton conference proceedings 2005* (pp. 201–210). European Federation of Explosives Engineers.

Cunningham, C. V. B. (2006). Concepts of blast hole pressure applied to blast design. *Fragblast, 10*(1–2), 33–45. https://doi.org/10.1080/13855140600852977

Cunningham, C. V. B. (2019). Keynote address: Optical fragmentation assessment—a technical challenge. In J. A. Franklin & T. Katsabanis (Eds.), *Measurement of blast fragmentation* (1st ed., pp. 13–19). Balkema. https://doi.org/10.1201/9780203747919-4

Da Gama, C. D. (1995). A model for rock mass fragmentation by blasting. In *8th ISRM congress* (pp. 73–76). International Society of Rock Mechanics.

Davies, P. A. (1995). Risk-based approach to setting of flyrock "danger zones" for blast sites. *Transactions—Institution of Mining & Metallurgy, Section A, 104*(May–August). https://doi.org/10.1016/0148-9062(95)99212-g

Deere, D. U., & Deere, D. W. (1989). *Rock Quality Designation (RQD) after twenty years* (Report GL-89-1, Vol. 53, Issue 9). Department of the Army, US Army Corps of Engineers. www.dtic.mil/cgi-bin/GetTRDoc?AD=ADA207597

Dick, R. D., Weaver, T. A., & Fourney, W. L. (1990). An alternative to cube-root scaling in crater analysis. In *Proceedings of the 3rd international symposium on rock fragmentation by blasting, Fragblast 3* (pp. 167–170). AusIMM.

Djordjevic, N. (2001). Practical application of detonation code for explosive selection. In *AusIMM proceedings, 306* (pp. 23–26). AusIMM.

Dobereiner, L., & de Freitas, M. H. (1986). Geotechnical properties of weak sandstones. *Geotechnique, 36*(1). https://doi.org/10.1680/geot.1986.36.1.79

Doucet, C., Cameron, A., & Lizotte, Y. (1996). Effects of rock mass characteristics on fragmentation in controlled blasting experiments in small development headings. In *Proceedings of the annual conference on explosives and blasting technique, 2* (pp. 1–10). International Society of Explosives Engineers.

Erkayaoglu, M., & Dessureault, S. (2019). Improving mine-to-mill by data warehousing and data mining. *International Journal of Mining, Reclamation and Environment, 33*(6), 409–424. https://doi.org/10.1080/17480930.2018.1496885

Esen, S., La Rosa, D., Dance, A., Valery, W., & Jankovic, A. (2007). Integration and optimisation of blasting and comminution processes. *Australasian Institute of Mining and Metallurgy Publication Series*, 95–103.

Esen, S., Onederra, I., & Bilgin, H. A. (2003). Modelling the size of the crushed zone around a blasthole. *International Journal of Rock Mechanics and Mining Sciences, 40*(4), 485–495. https://doi.org/10.1016/S1365-1609(03)00018-2

Fang, Q., Nguyen, H., Bui, X. N., Nguyen-Thoi, T., & Zhou, J. (2021). Modeling of rock fragmentation by firefly optimization algorithm and boosted generalized additive model. *Neural Computing and Applications, 33*(8), 3503–3519. https://doi.org/10.1007/s00521-020-05197-8

Gao, W., Alqahtani, A. S., Mubarakali, A., Mavaluru, D., & Khalafi, S. (2020). Developing an innovative soft computing scheme for prediction of air overpressure resulting from mine blasting using GMDH optimized by GA. *Engineering with Computers, 36*(2), 647–654. https://doi.org/10.1007/s00366-019-00720-5

Ghose, A. K. (1988). Design of drilling and blasting subsystems—a rock mass classification approach. *Proceedings of the Symposium on Mine Planning and Equipment Selection*, 335–340. https://doi.org/10.1016/0148-9062(90)90151-q

Gilbride, L., Taylor, S., Songlin Zhang, Daemen, J. J. K., & Mousset- Jones, P. (1995). Blast-induced rock movement modelling for Nevada gold mines. *Mineral Resources Engineering, 4*(2), 175–193. https://doi.org/10.1142/S0950609895000175

Grundstrom, C., Kanchibotla, S. S., Jankovich, A., & Thornton, D. (2001). Blast fragmentation for maximizing the sag mill throughput at Porgera gold mine. In *Proceedings of the annual conference on explosives and blasting technique, 1* (pp. 383–399). International Society of Explosives Engineers.

Hagan, T. N., & Just, G. D. (1974). Rock breakage by explosives - theory, practice and optimization. In *Proceedings of the 3rd ISRM congress* (pp. 1349–1358). International Society of Rock Mechanics.

Hasanipanah, M., & Amnieh, H. B. (2020). A fuzzy rule-based approach to address uncertainty in risk assessment and prediction of blast-induced flyrock in a quarry. *Natural Resources Research, 29*(2), 669–689. https://doi.org/10.1007/s11053-020-09616-4

Hustrulid, W. (1999a). *Blasting principles for open pit mining: Volume 1—general design concepts*. Balkema.

Hustrulid, W. (1999b). *Blasting principles for open pit mining: Volume 2—theoretical foundations*. Balkema.

Ibarra, J. A., Maerz, N. H., & Franklin, J. A. (1996). Overbreak and underbreak in underground openings part 2: Causes and implications. *Geotechnical and Geological Engineering, 14*(4). https://doi.org/10.1007/BF00421947

Inanloo, A. S. H., Sereshki, F., Ataei, M., & Karamoozian, M. (2018). Investigation of rock blast fragmentation based on specific explosive energy and in-situ block size. *International Journal of Mining and Geo-Engineering, 52*(1), 2–7.

Innaurato, N., Mancini, R., & Cardu, M. (1998). On the influence of rock mass quality on the quality of blasting work in tunnel driving. *Tunnelling and Underground Space Technology, 13*(1), 81–89. https://doi.org/10.1016/S0886-7798(98)00027-3

ISEE (2022) International Society of Explosive Engineers, online database (http://www.isee.org), members only portal.

Jang, H., & Topal, E. (2013). Optimizing overbreak prediction based on geological parameters comparing multiple regression analysis and artificial neural network. *Tunnelling and Underground Space Technology*, *38*, 161–169. https://doi.org/10.1016/j.tust.2013.06.003

Jankovic, A., & Valery, W. (2002). Mine-to-mill optimisation for conventional grinding circuits—a scoping study. *Journal of Mining and Metallurgy, Section A: Mining*, *38*(1–4), 49–66.

Jeng, F. S., Weng, M. C., Lin, M. L., & Huang, T. H. (2004). Influence of petrographic parameters on geotechnical properties of tertiary sandstones from Taiwan. *Engineering Geology*, *73*(1–2), 71–91. https://doi.org/10.1016/j.enggeo.2003.12.001

Kamani, M., & Ajalloeian, R. (2019). Evaluation of engineering properties of some carbonate rocks trough corrected texture coefficient. *Geotechnical and Geological Engineering*, *37*(2). https://doi.org/10.1007/s10706-018-0630-8

Khademian, A., & Bagherpour, R. (2017). Environmentally sustainable mining through proper selection of explosives in blasting operation. *Environmental Earth Sciences*, *76*(4). https://doi.org/10.1007/s12665-017-6483-2

Kiliç, A. M., Yaşar, E., Erdoğan, Y., & Ranjith, P. G. (2009). Influence of rock mass properties on blasting efficiency. *Scientific Research and Essays*, *4*(11), 1213–1224.

Konya, C. J. (1995). *Blast design* (1st ed.). Intercontinental Development Corporation. https://books.google.co.in/books?id=AdcOAQAAMAAJ

Konya, C. J., & Walter, E. J. (1991). *Rock blasting and overbreak control* (No. FHWA-HI-92–001; NHI-13211). National Highway Institute, 430p.

Koopialipoor, M., Armaghani, D. J., Haghighi, M., & Ghaleini, E. N. (2019). A neuro-genetic predictive model to approximate overbreak induced by drilling and blasting operation in tunnels. *Bulletin of Engineering Geology and the Environment*, *78*(2), 981–990. https://doi.org/10.1007/s10064-017-1116-2

Kou, S. Q., & Rustan, A. (1992). Burden related to blasthole diameter in rock blasting. *International Journal of Rock Mechanics and Mining Sciences*, *29*(6), 543–553. https://doi.org/10.1016/0148-9062(92)91612-9

Laubscher, D. H., & Jakubec, J. (2001). The MR rock mass classification for jointed rock masses. In W. A. Hustrulid & R. L. Bullock (Eds.), *Underground mining methods: Engineering fundamentals and international case studies* (pp. 475–481). Society of Mining Metallurgy and Exploration.

Leng, Z., Fan, Y., Gao, Q., & Hu, Y. (2020). Evaluation and optimization of blasting approaches to reducing oversize boulders and toes in open-pit mine. *International Journal of Mining Science and Technology*, *30*(3), 373–380. https://doi.org/10.1016/j.ijmst.2020.03.010

Li, N. N., Li, J. C., Li, H. B., & Chai, S. B. (2016). Experimental study on dynamic response of rock masses with different joint matching coefficients. In *Rock dynamics: From research to engineering—2nd international conference on rock dynamics and applications, ROCDYN 2016* (pp. 129–134). International Society of Rock Mechanics. https://doi.org/10.1201/b21378-18

Lilly, P. A. (1986). Empirical method of assessing rock mass blastability. *Symposia Series—Australasian Institute of Mining and Metallurgy*, 89–92. https://doi.org/10.1016/0148-9062(87)92504-6

Livingston, C. W. (1956). Fundamentals of rock failure. *Quarterly of the Colorado School of Mines*, *51*(3), 1–11.

Long, Y., Zhong, M. S., Xie, Q. M., Li, X. H., Song, K. J., & Liao, K. (2013). Influence of initiation point position on fragmentation by blasting in iron ore. In *Rock fragmentation by blasting, FRAGBLAST 10—Proceedings of the 10th international symposium on rock fragmentation by blasting* (pp. 111–116). CRC Press/Balkema. https://doi.org/10.1201/b13759-18

Lu, P., & Latham, J. P. (1998). A model for the transition of block sizes during fragmentation blasting of rock masses. *Fragblast*, *2*(3), 341–368. https://doi.org/10.1080/13855149809408781

Mackenzie, A. (1966). Cost of explosives—do you evaluate it properly? *Mining Congress Journal*, *52*(5), 32–41.

Marinos, V., Marinos, P., & Hoek, E. (2005). The geological strength index: Applications and limitations. *Bulletin of Engineering Geology and the Environment*, *64*(1), 55–65. https://doi.org/10.1007/s10064-004-0270-5

McKenzie, C. K., Stacey, G. P., & Gladwin, M. T. (1982). Ultrasonic characteristics of a rock mass. *International Journal of Rock Mechanics and Mining Sciences*, *19*(1), 25–30. https://doi.org/10.1016/0148-9062(82)90707-0

Mertuszka, P., Cenian, B., Kramarczyk, B., & Pytel, W. (2018). Influence of explosive charge diameter on the detonation velocity based on Emulinit 7L and 8L Bulk emulsion explosives. *Central European Journal of Energetic Materials*, *15*(2), 351–363. https://doi.org/10.22211/cejem/78090

Messaoud, S. B., Hamdi, E., & Gaied, M. (2020). Coupled geomechanical classification and multivariate statistical analysis approach for the optimization of blasting rock boulders. *Arabian Journal of Geosciences*, *13*(15), 1–19. https://doi.org/10.1007/s12517-020-05604-3

Michaux, S., & Djordjevic, N. (2005). Influence of explosive energy on the strength of the rock fragments and SAG mill throughput. *Minerals Engineering*, *18*(4), 439–448. https://doi.org/10.1016/j.mineng.2004.07.003

Mishra, A. K., & Sinha, M. (2016). Influence of bulk emulsion explosives property on fragmentation and productivity in an opencast coal mine. *Journal of Mines, Metals and Fuels*, *64*(7), 303–309.

Monjezi, M., Bahrami, A., Varjani, A. Y., & Sayadi, A. R. (2011). Prediction and controlling of flyrock in blasting operation using artificial neural network. *Arabian Journal of Geosciences*, *4*(3–4), 421–425. https://doi.org/10.1007/s12517-009-0091-8

Monjezi, M., Dehghani, H., Singh, T. N., Sayadi, A. R., & Gholinejad, A. (2012). Application of TOPSIS method for selecting the most appropriate blast design. *Arabian Journal of Geosciences*, *5*(1), 95–101. https://doi.org/10.1007/s12517-010-0133-2

Morrell, S., & Munro, P. D. (2000). Increased profits through mine-and-mill integration. *Australasian Institute of Mining and Metallurgy Publication Series*, *2*, 194–198.

Mortazavi, A., & Katsabanis, P. D. (2000). Modelling the effects of discontinuity orientation, continuity, and dip on the process of burden breakage in bench blasting. *Fragblast*, *4*(3–4), 175–197. https://doi.org/10.1076/frag.4.3.175.7390

Muftuoglu, Y. V., Amehmetoglu, A. G., & Karpuz, C. (1991). Correlation of powder factor with physical rock properties and rotary drill performance in Turkish surface coal mines. In *Proceedings of the 7th ISRM congress*. International Society of Rock Mechanics.

Nageshwaraniyer, S. S., Kim, K., & Son, Y. J. (2018). A mine-to-mill economic analysis model and spectral imaging-based tracking system for a copper mine. *Journal of the Southern African Institute of Mining and Metallurgy*, *118*(1), 7–14. https://doi.org/10.17159/2411-9717/2018/v118n1a2

Ning, Y., Yang, J., Ma, G., & Chen, P. (2011). Modelling rock blasting considering explosion gas penetration using discontinuous deformation analysis. *Rock Mechanics and Rock Engineering*, *44*(4), 483–490. https://doi.org/10.1007/s00603-010-0132-3

Nur Lyana, K., Hareyani, Z., Kamar Shah, A., & Mohd. Hazizan, M. H. (2016). Effect of geological condition on degree of fragmentation in a Simpang Pulai marble quarry. *Procedia Chemistry*, *19*, 694–701. https://doi.org/10.1016/j.proche.2016.03.072

Ouchterlony, F. (2005). The Swebrec© function: Linking fragmentation by blasting and crushing. *Institution of Mining and Metallurgy. Transactions. Section A: Mining Technology*, *114*(1). https://doi.org/10.1179/037178405X44539

Ouchterlony, F., Nyberg, U., Mats, O., Ingvar, B., Lars, G., & Henrik, G. (2004). Where does the explosive energy in rock blasting rounds go? *Science and Technology of Energetic Materials*, *65*(2), 54–63.

Palmström, A. (1996). RMi—a system for characterizing rock mass strength for use in rock engineering. *Journal of Rock Mechanics and Tunnelling Technology*, *1*(2), 1–40. www.rockmass.net/ap/46_Palmstrom_on_RMi_for_rockmass_strength.pdf

Park, J., & Kim, K. (2020). Use of drilling performance to improve rock-breakage efficiencies: A part of mine-to-mill optimization studies in a hard-rock mine. *International Journal of Mining Science and Technology*, *30*(2), 179–188. https://doi.org/10.1016/j.ijmst.2019.12.021

Qu, S., Hao, S., Chen, G., Li, B., & Bian, G. (2002). The BLAST-CODE model–a computer-aided bench blast design and simulation system. *Fragblast*, *6*(1), 85–103.

Raina, A. K. (2013). Blast fragmentation assessment and optimization: Back to basics. *Journal of Mines, Metals and Fuels*, *61*(7–8), 207–212.

Raina, A. K. (2014). *Modelling the flyrock in opencast blasting under difficult geomining conditions*. Ph.D. Thesis, Indian Institute of Technology—ISM. www.iitism.ac.in/pdfs/departments/mining/Research-degrees-completed.pdf

Raina, A. K., & Trivedi, R. (2019). Exploring rock—explosive interaction through cross blast-hole pressure measurements. *Geotechnical and Geological Engineering*, *37*(2), 651–658. https://doi.org/10.1007/s10706-018-0635-3

Rakishev, B. R. (1981). A new characteristic of the blastability of rock in quarries. *Soviet Mining Science*, *17*(3), 248–251. https://doi.org/10.1007/BF02497198

Rosin, P., & Rammler, E. (1933). The laws governing the fineness of powdered coal. *Journal of the Institute of Fuel*, *7*, 29–36.

Rustan, A., Cunningham, C. V. B., Fourney, W., & Spathis, A. (2011). Mining and rock construction technology desk reference. In A. Rustan (Ed.), *Mining and rock construction technology desk reference*. CRC Press/Balkema. https://doi.org/10.1201/b10543

Salmi, E. F., & Sellers, E. (2021). A review of the methods to incorporate the geological and geotechnical characteristics of rock masses in blastability assessments for selective blast design. *Engineering Geology*, *281*. https://doi.org/10.1016/j.enggeo.2020.105970

Sanchidrián, J. A., Segarra, P., & López, L. M. (2007). Energy components in rock blasting. *International Journal of Rock Mechanics and Mining Sciences*, *44*(1), 130–147. https://doi.org/10.1016/j.ijrmms.2006.05.002

Sayevand, K., & Arab, H. (2019). A fresh view on particle swarm optimization to develop a precise model for predicting rock fragmentation. *Engineering Computations (Swansea, Wales)*, *36*(2), 533–550. https://doi.org/10.1108/EC-06-2018-0253

Scott, A. (2020). 'Blastability' and blast design. In *Rock Fragmentation by Blasting*, Mohanty B. (Ed.), CRC Press, London, pp. 27–36. https://doi.org/10.1201/9781003078104

Scott, A., Morrell, S., & Clark, D. (2002). Tracking and quantifying value from "mine to mill" improvement. *Australasian Institute of Mining and Metallurgy Publication Series*, *8*, 77–84.

Scott, A., & Onederra, I. (2015). Characterising the Blasting Properties of Iron Ore. *Proceedings Iron Ore 2015: Maximising Productivity*, Australasian Institute of Mining and Metallurgy, (pp. 13–15).

Segaetsho, G., & Zvarivadza, T. (2019). Application of rock mass classification and Blastability Index for the improvement of wall control: A hardrock mining case study. *Journal of the Southern African Institute of Mining and Metallurgy*, *119*(1), 31–40. https://doi.org/10.17159/2411-9717/2019/v119n1a4

Sellers, E., Salmi, E. F., Usami, K., Greyvensteyn, I., & Mousavi, A. (2019). Detailed rock mass characterization—a prerequisite for successful differential blast design. *5th ISRM*

Young Scholars' Symposium on Rock Mechanics and International Symposium on Rock Engineering for Innovative Future, YSRM 2019, pp. 577–582.

Sim, Y., Cho, G. C., & Song, K. II. (2017). Prediction of fragmentation zone induced by blasting in rock. *Rock Mechanics and Rock Engineering*, *50*(8), 2177–2192. https://doi.org/10.1007/s00603-017-1210-6

Singh, S. P. (2003). The influence of rock mass quality in controlled blasting. *Australasian Institute of Mining and Metallurgy Publication Series*, *1*, 219–222.

Thornton, D., Kanchibotla, S. S., & Brunton, I. (2002). Modelling the impact of rockmass and blast design variation on blast fragmentation. *Fragblast*, *6*(2), 169–188. https://doi.org/10.1076/frag.6.2.169.8663

Tsiambaos, G., & Saroglou, H. (2010). Excavatability assessment of rock masses using the Geological Strength Index (GSI). *Bulletin of Engineering Geology and the Environment*, *69*(1), 13–27. https://doi.org/10.1007/s10064-009-0235-9

Tsidzi, K. E. N. (1990). The influence of foliation on point load strength anisotropy of foliated rocks. *Engineering Geology*, *29*(1), 49–58. https://doi.org/10.1016/0013-7952(90)90081-B

Williamson, S. R., & Armstrong, M. E. (1986). Measurement of explosive product gas penetration. *Symposia Series—Australasian Institute of Mining and Metallurgy*, 147–152. https://doi.org/10.1016/0148-9062(88)92975-0

Wu, M., Cai, Q., & Shang, T. (2019). Assessing the suitability of imperialist competitive algorithm for the predicting aims: An engineering case. *Engineering with Computers*, *35*(2), 627–636. https://doi.org/10.1007/s00366-018-0621-7

Yi, C., Johansson, D., & Greberg, J. (2018). Effects of in-situ stresses on the fracturing of rock by blasting. *Computers and Geotechnics*, *104*, 321–330. https://doi.org/10.1016/j.compgeo.2017.12.004

Zhang, K. (1990). On the geomechanics of rock mass blasting and blastibility classification. *Scientia Geologica Sinica*, *2*, 194–199.

Zhang, Z., & Luukkanen, S. (2021, April). Feasibility and necessity of mine to mill optimization in mining industry. *Materia Medica*, *2*, 63–66.

Zhang, Z.-X. (2016). *Rock fracture and blasting: Theory and applications*. Butterworth-Heinemann.

Zou, D. (2016). *Theory and technology of rock excavation for civil engineering*. Springer Singapore, https://doi.org/10.1007/978-981-10-1989-0

2 Flyrock! What is It?

Flyrock, one of the undesired outcomes of rock blasting with a fair idea about the phenomenon, was introduced in Chapter 1. When one talks of flyrock in general sense, the term "flyrock" involves the whole phenomenon of flyrock that has several components that fall in different physical and analytical domains of complex nature. However, there are different aspects of flyrock when discussing it in specific terms. Accordingly, it is important to have a critical understanding of the phenomenon, the definitions of flyrock as used in various reports and publications, the intricacies of the terminology and the confusions thereof, and its occurrence in various types of excavations. While the term flyrock has been associated with blasting only, more recently, it has also been used in a different context of rockfall. Such usage has further complicated the definition of flyrock. As such, it is important to dispel the truth about flyrock (discussed in detail in Chapter 3), and at the same time put all definitions and domains into proper perspective so that the phenomenon can be understood in totality.

A simple representation of the flyrock phenomenon is provided in Figure 2.1, wherein it can be seen that a rock fragment, detached from the main muck of the blast, is moving under the impact of high-pressure explosive gases, and is thrown farther than the distance planned for normal throw of a blast. Such flyrock can emerge from any part of the face or bench being blasted, can be launched at any angle, and can travel in any direction in 360° realm, and in all the three dimensions.

A proper distinction between different elements, which are relevant to blasting or flyrock, has been constructed in Figure 2.1, while appreciating that there are different

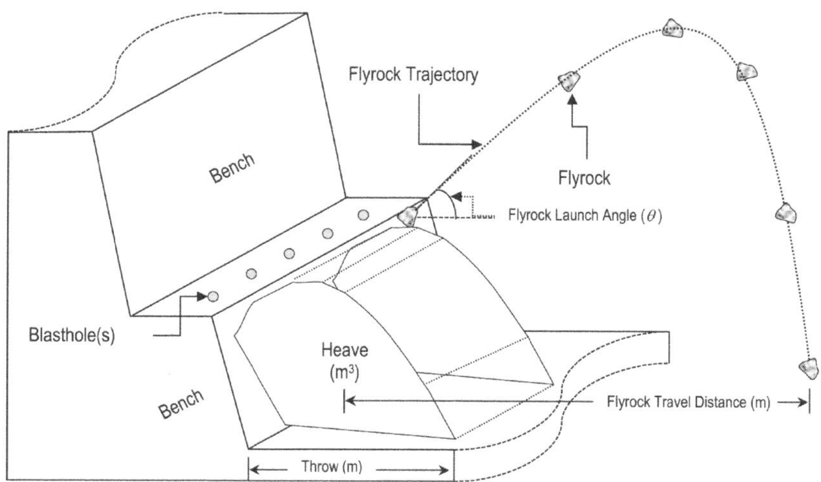

FIGURE 2.1 Flyrock as a function of blast and comparison with the throw of the blast.

DOI: 10.1201/9781003327653-2

aspects of the movement of the fragmented rockmass that converts to heaved muck and a fragment under the influence of explosive gas pressure. Accordingly, the definition of flyrock should not be so complicated and ambiguous as demonstrated further.

2.1 DEFINITIONS AND CONFUSION IN TERMINOLOGY AND REPORTING

Flyrock has been used in published literature for different events that may or may not belong to blasting. There is evidence that scientists and engineers have used multiple terminologies to portray flyrock. Despite the efforts made by ISRM to standardize the nomenclature of blasting (Rustan et al., 2011), scholars prefer to use terminologies, symbols, acronyms, and units of their choice, probably for the sake of their convenience or reasons better known to them. The case with flyrock is no different. One can find the use of terms "flyrock" (correct form used by a lot of authors), "fly rock or fly rocks" (Dhekne, 2015; Frasher & Dambov, 2018; Shevkun et al., 2018), "fly-rock" (Guo et al., 2019, 2021; Benwei et al., 2016; Tao et al., 2018), "flying stones" (Huang et al., 2020; Qiao, 2020; Zijun, 2002), and "blasting flying stones" (Yao et al., 2011), in various texts, which has resulted in ambiguity.

In order to demonstrate the aforementioned inference, a summary of the number of citations using the terms "flyrock," "fly-rock" or "fly rock," and "flying stones" in the title or abstract of publications pertaining to blasting, discovered at different online resources, is provided in Table 2.1.

The variation in use of terminology for flyrock is amply evident from the data provided in Table 2.1. A general text search for the aforementioned terms yields different results in general scientific search engines. Similarly, most significant works on flyrock cite blast design variables, factors, and rockmass characteristics by self-devised symbols and short forms without a firm scientific basis. Also, flyrock has been defined in a diverse manner, most of which cannot be regarded as scientific. This is the reason for confusion and makes it difficult for a researcher to search and acquire the literature. Hence, it is important to dispel the doubts in flyrock nomenclature while introducing the concept of the flyrock. A need exists for an effort by journals and editors to investigate this aspect of standardization of nomenclature.

2.1.1 Flyrock Definition

As mentioned earlier, flyrock is a standard term used for one or multiple fragments of rock that eject from a blast face and travel beyond the desired or stipulated distance. It must be remembered that flyrock is a rock fragment and does not qualify the dimensional and vector attributes. Even the use of "flyrock fragment(s)" (Sawmliana et al., 2020; Stojadinović et al., 2011) do not qualify the definition, as flyrock in itself denotes a rock fragment. A host of definitions of flyrock exist in reports and a few are compiled in Table 2.2, along with brief observations.

There are some definitions in Table 2.2 which deal with the cause of flyrock, while others refer to the effect of flyrock, and yet others have inherent anomalies, including verbal issues, while defining flyrock.

TABLE 2.1
Citation Count of "Flyrock," "Fly Rock" or "Fly-rock," and "Flying Stones" Search in Different Online Databases as of June 2022

Organization/Society/ Search Portal▶	ISEE	AusIMM	NIOSH	CIM	SME-ICGCM	SME	AIME	SAIMM	IOM3	Google Scholar	Mendeley
Search criterion▼						**Citation Count**					
Flyrock—all in title	87	22	12	–	–	7	2	2	1	228	244
Fly rock or fly-rock—all in title	28	3	7	3	1	8	4	3	–	67	157
Flying stones (for flyrock only)—all in title or abstract	1	–	–	–	–	1	–	–	–	4	5

Note: ISEE—International Society of Explosive Engineers, AusIMM—Australasian Institute of Mining and Metallurgy, NIOSH—National Institute for Occupational Safety and Health, CIM—Canadian Institute of Mining Metallurgy and Petroleum, SME—Society for Mining Metallurgy and Exploration, ICGCM—International Conference on Ground Control in Mining, AIME—The American Institute of Mining, Metallurgical, and Petroleum Engineers, SAIMM—Southern African Institute of Mining and Metallurgy, IOM3—Institute of Materials Mineral and Mining—accessed through ISEE OneMine portal (members-only portal).

TABLE 2.2
Definitions of Flyrock as Used by Different Researchers

S. No.	Author(s)	Definition	Comments
1	Davies (1995)	"Flyrock, is essentially the product of uncontrolled venting of the gases that are developed when a charge is initiated"	Explains mechanism of flyrock but does not give a proper definition
2	Dick et al. (1983); Fletcher and D'Andrea (1990)	"Flyrock is the rock that is propelled through the air during mine blasting"	Is a good yet not complete definition of flyrock as the term rock in the context of flyrock should have been used as a fragment
3	Persson et al. (1994)	Flyrock(s) are the "Rock fragments thrown unpredictably from a blasting site by the force of explosion"	Is a good definition
4	Siskind and Kopp (1995)	"Flyrock can be defined as an undesirable throw of the material"	Throw and material are confusing terms as these can be anything and the blast term is missing
5	IME (1997)	"Flyrock is defined as the rock propelled beyond the blast area by the force of an explosion"	This is a proper definition
6	Kane (1997)	"Flyrock also called rock throw, is uncontrolled propelling of rock fragments produced in blasting and constitutes one of the main sources of material damage and harm to people"	This is confusing as there is a lot of difference in throw and flyrock. Focuses on effect
7	Gibson and St George (2001)	Flyrock can be described essentially as the unexpected and undesirable projection of rock fragments during detonation of a blast	
8	Little (2007)	"Flyrock—the undesired propulsion of rock fragments through the air or along the ground beyond the blast zone by the force of the explosion that is contained within the blast clearance (exclusion) zone"	Qualifies the flyrock with respect to its desirability and relates it to blast danger zone. The "propulsion of rock fragments" cannot be a flyrock
9	Raina and Murthy (2016)	"Flyrock is a rock fragment that travels excessive and unwanted distances from a blast face in surface blasting under the impulse of explosive gases"	Gives a broad context of the definition

S. No.	Author(s)	Definition	Comments
10	Lwin and Aung (2019)	Flyrock refers to the uncontrolled dispersion of rock fragments from blast areas caused by explosive energy	"Uncontrolled dispersion of rock fragments" is not a correct definition
11	Office of the Revisor of Statutes, Maine[1]	"Flyrock means rock that is propelled through the air or across the ground as a result of blasting and that leaves the blast area"	This is a good definition provided blast area is defined properly
12	Han et al. (2020)	"Any blasting energy in mining and civil engineering projects produce a sudden ejection of rock pieces, which are referred to as flyrock"	The definition "a sudden ejection of rock pieces" is unclear
13	Zhou et al. (2020)	"Flyrock is one of the environmental problems defined as the throw or move of rock fragments due to excessive pressure caused by an unexpected blast of explosives"	Throw or movement cannot be a flyrock, hence the definition is not correct
14	Hosseini et al. (2022)	"Flyrock is determined as the rock fragments propelled beyond the blasting region by the energy of charges"	The definition is better but not properly composed and has verbal issues

Note: The list is not exhaustive.

The ISRM recognize the following definition of flyrock provided by Rustan et al. (2011) as:

> *Any rock fragments thrown unpredictable from a blasting site. Flyrock may develop in the following situations: insufficient stemming, too small a burden because of overbreak from the preceding or proceeding blast, planes of weakness in rock which reduce the resistance to blasting, and finally existence of loose rock fragments on top of the bench.*

Although this definition is not complete, as will be further elaborated in Chapter 4, it dispels the notion that flyrock is a fragment. It also raises doubt about the prediction and the causes mentioned that are not inclusive. There may be several other interpretations like this, but a simple definition of flyrock could be "any rock fragment or fragments travelling beyond desired distance from a blast due to impulse of explosive gases." Use of special rockmass conditions, over-charging explosive, extreme blast design, or a combination of all these can be attempted to be part of the definition of the flyrock.

There is even mention of "excessive or excess flyrock" in some texts (Fletcher & D'Andrea, 1990; Kopp, 1994; Workman & Calder, 1994) that appear to be unnecessary on account of the aforementioned definition of flyrock. Such insertions can be best avoided and included in the term "flyrock" itself. A reason for use of excessive flyrock can possibly be due to more than one flyrock, which can also come under the ambit of its definition.

The following inferences are thus derived from the said discussion:

1. The definition of flyrock as propounded by Rustan et al. (2011) should be uniformly adopted by all concerned. Further modifications can be suggested to ISRM for modification of the definition.
2. The term "flyrock" should be used in reporting, and other terminologies that do not conform to the standard definitions, as cited earlier, should be discarded.
3. The questions as to what was the size of the flyrock and what distance did it travel should be used to qualify a flyrock.
4. The cause of flyrock, the size of flyrock, the impulse with which it is launched, its travel in air, and the impact on any object it lands on are a different proposition. As such, these attributes of flyrock should be excluded from its definition.
5. The uniformity in terminology and definition of flyrock will be an advantage to scholars and writers in their future research on the subject and will facilitate standardization.

In view of this discussion, a suggested definition of flyrock is as follows:

A rock fragment that ejects from a blast face, under the influence of high-pressure impact of explosive gases, the occurrence of which can be defined in terms of its probabilities, and can travel distances more than several bench heights, while impacting the objects of concern near the mines or blasting and has a potential to damage and cause fatalities.

Since distance of travel of a flyrock is the main concern, rock fragments falling a few bench heights or beyond a couple of bench widths from the blast can be considered as flyrock and for that reason, the aforementioned definition can thus be refined.

2.1.2 Flyrock and Associated Terminology

Several terms used while discussing flyrock are provided here for the sake of clarity of the definitions and to remove confusion during expressions for different aspects of flyrock:

- Flyrock should be defined as suggested earlier.
- Throw is "the movement of the material disintegrated and fragmented in blasting" (Rustan et al., 2011) and is expressed as throw length (R_T) and termed as "throw" hereon. Throw is a regular effect of a blast that can extend up to a few bench heights from the blast location and is a desired outcome. Due to the dynamic nature of blasting, it is difficult to conduct measurements during blasting. However, post-blast investigations on the

extent of the muck and its heaving can be easily done with modern methods of 3D mapping using laser-based and GPS technologies. There are several instances of such data that have been correlated with the blast design and its other variables, including explosive characteristics (Choudhary & Rai, 2013; Sharma et al., 2019; Xiao et al., 2019). However, a need exists to distinguish throw of muck from throw of flyrock, which can be designated as flyrock travel distance as there is a difference in prediction mechanisms of both the occurrences. Hence, for the purpose of blasting, use of the term "flyrock throw" can be avoided.

- Flyrock size (k_f) is the size of the flyrock that has been projected beyond the normal throw of the blast. The k_f does not find mention at most of the places, but it is an important aspect of flyrock because the damaging effect will be more by larger fragments, in case these hit an object. Also, the travel history of flyrock will be determined by its size. However, there is a discrepancy in the definition of the k_f owing to the irregular nature of such rock fragments from blasting and which dimension should be considered for analysis.
- Flyrock launch or exit velocity (v_f or v_0) is the velocity of the flyrock at the time of its ejection from the blast face. Such velocity has also been called the initial or exit velocity and needs high-speed videography to measure it. There have been attempts to quantify it, but the method of 2D photogrammetry used in such calculations has its limitations. The reliability of such measurements needs to be established. In trajectory physics, v_f has a significant role as it determines the travel distance of a flyrock to a major extent.
- Flyrock launch angle (θ) is the angle at which a flyrock is ejected or launched with respect to the horizontal. The launch angle of flyrock can be easily determined with careful planning through 2D photogrammetry as it is scale-independent.
- Flyrock distance or range (R_f) is the horizontal distance that a flyrock travels from its origin in a particular direction. The "distance" is the length of the space between two points that is qualified by the "horizontal" to match the meaning of the flyrock distance. The definition of flyrock distance has been misread as flyrock in several instances and hence a distinction is thought to be necessary.
- Flyrock trajectory is the path along which a flyrock has travelled under the impact of high-pressure gases. The definition of "trajectory" in Oxford Dictionary, "*the path followed by a projectile flying or an object moving under the action of given forces*," qualifies best for the definition of flyrock trajectory. The length of the path should never be mistaken for flyrock distance.

The definitions of independent terms here are provided to remove the confusion existing in the literature regarding various terms used in flyrock studies. The clarity in definitions is expected to help future researchers to include such terminology and symbols in a scientific and candid manner.

2.2 THE CONCEPT OF WILD FLYROCK

The concept of wild flyrock finds its initial mention in Roth (1979) who dedicated a full section to the topic. He defines the wild flyrock as follows:

> *Wild flyrock may be defined as flyrock that travels much further than flyrock that is normally encountered in any given blasting operation or further than estimated by existing rules of thumb.*

Little (2007) also refers to wild flyrock as the unexpected propulsion of rock fragments owing to odd blast or rockmass condition. A flyrock is wild if it clears the blast danger zone. Although the terminology used is not proper, the author probably tried to distinguish between flyrock travelling few tens of metres from few hundred metres (distinction is not clear from their report). Moore and Richards (2011) also mention that the wild flyrock poses a threat to the mine infrastructure. They, however, do not refer to wild flyrock further but focus on flyrock distance prediction. It appears that the concept was developed to include all uncertainties in rockmass, blast design, and initiation sequencing or failures. Significant instances of such irregularities that can result in flyrock have been logged by Roth (1979).

A conceptual diagram of distinction between wild and general flyrock is shown in Figure 2.2 as conceived from the definitions of a few authors. However, it is difficult

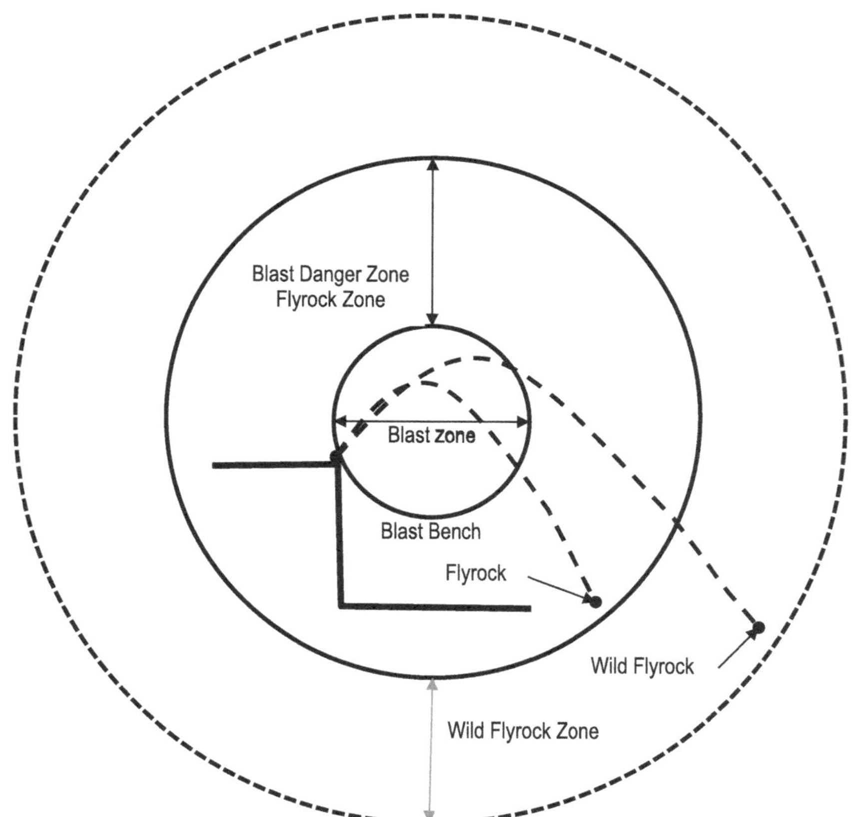

FIGURE 2.2 The concept of wild flyrock as explained by various authors.

to draw the distinction between the two types of flyrock(s) mentioned. Even if a flyrock is considered to fall in the BDZ and a wild flyrock is deemed to have crossed the BDZ, the difference is not clear as the BDZ is not scientifically defined yet, as discussed in Chapter 6. There is a possibility of specifying these based on distance of their travel. Nonetheless, that does not help to distinguish between the two terms: flyrock and wild flyrock.

It may be pointed out that the definition of wild flyrock is unclear, since by the definition used by a few authors, the wild flyrock is practically "uncontrolled rock fragment emanating from a blast" that travels several hundred metres under the influence of explosive pressures and from reasons practically not in control of the blasting crew and that implications of the wild flyrock can be very dangerous is perfectly matching the definition of flyrock. Also, the term "wild flyrock" has been ambiguated with general definition of flyrock[2] and the readers must be cautioned about the same. The data on such flyrock is scant (Roth, 1979) and hence it is better to include all flyrock(s) under one definition.

2.3 THE REBOUND EFFECT

A flyrock once generated lands at a specific location that is determined by the impact, and hence its initial velocity (v_0) and its launch angle (θ) under the influence of air drag (defined in Chapter 3). Almost all the models that predict flyrock travel distance or flyrock distance focus only on this part of the distance. However, there is another component to the flyrock that adds to its total travel distance. This is called the rebound effect (Raina et al., 2013, 2015). The phenomenon is explained further with the help of Figure 2.3.

The concept, as explained in Figure 2.3, is that once a flyrock lands on its final surface like bench or any other object, most of the energy it is carrying is absorbed by the surface. However, owing to the remanent energy, it tends to rebound, not once, but several times, depending on the surface of landing. The forces acting on the flyrock during rebound may even exert a torque on the flyrock. The total flyrock distance will then be different from that assumed by the trajectory equations, as presented in the case studies by various researchers. This may be one of the reasons that correlations of flyrock distance are not as expected with the variables defining its travel.

Another special condition that can be visualized under this situation is further explained by Figure 2.4.

In this case, the flyrock is generated from a top bench blast and projected to the bottom of the bench. Hereon, the flyrock is further reflected from the bench surface to a lower bench and so on. The total travel distance in this case will be quite different from the assumed distance. A similar condition, as described earlier, was simulated by Anh Tuan (2020) using non-smooth discrete element method for modelling of flyrock distance while incorporating the rebound of fragments with arbitrary slopes and rock geometries. The author provided code for determining the influence of shape of fragments on rolling and determined the flyrock initial velocities as a function of blast design variables like specific charge, charge per hole, and blasthole length. Anh Tuan (2020) determined that the coefficient of restitution, i.e. the ratio of final velocity to the initial velocity, was equal to 0.6.

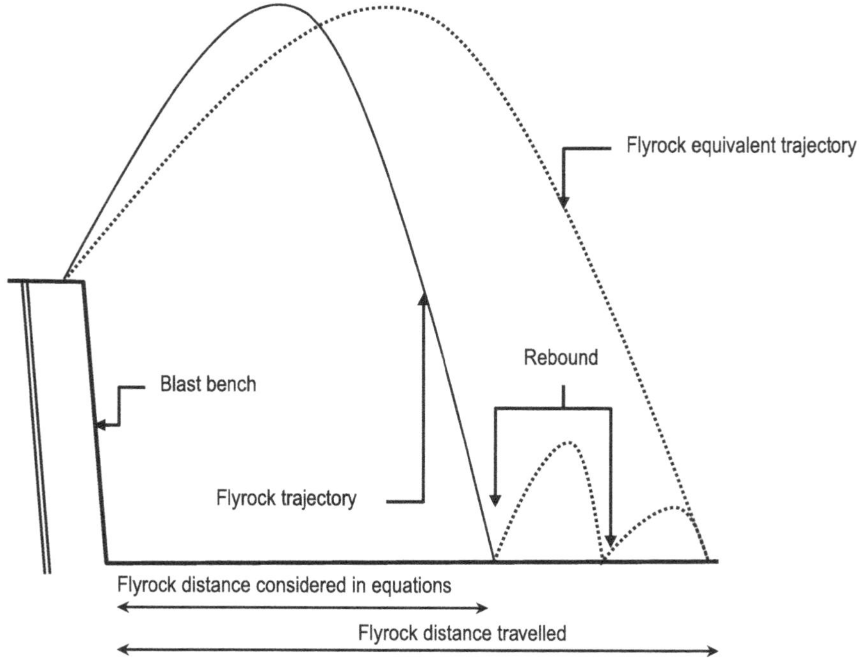

FIGURE 2.3 The concept of rebound of flyrock and its influence on its total travel distance.

Experimental evidence of the rebound was generated with the help of blasting in concrete models by the author (Raina et al., 2006). A total of 67 concrete models with different configurations were blasted. Complete description of such blasts will be provided in Chapter 3 along with further analysis. The distances of flyrock measured manually and calculated with the help of high-speed video of the blasts did not match. Rebound of flyrock could be recorded in 34 models that accounted for almost 50% of the trials. The details of the flyrock rebound velocities and rebound angles thus obtained are provided in Figures 2.5 and 2.6, respectively.

The ratio of rebound velocity to the initial velocity of the flyrock given in Figure 2.7 points to the fact that there is significant residual energy, in terms of initial velocity, in the flyrock and lies between 0.14 and 0.74 of the original velocity. Most of such velocity lies in the range of 0.29–0.44 of the initial velocity.

The frequency of ratio of rebound angle to the initial launch angle (degrees) of the flyrock is shown in Figure 2.8. The histogram points to the fact that the angle of rebound in most of the cases is less than initial launch angle, though there are instances where such angle can be higher than the initial launch angle.

The flyrock distances calculated in the aforementioned tests in concrete blocks did not account fully for the measured distances, despite adding the rebound distance. The problem was further investigated with the help of high-speed video analysis, and it was discovered that the flyrock not rebounding on the surface was further dragged

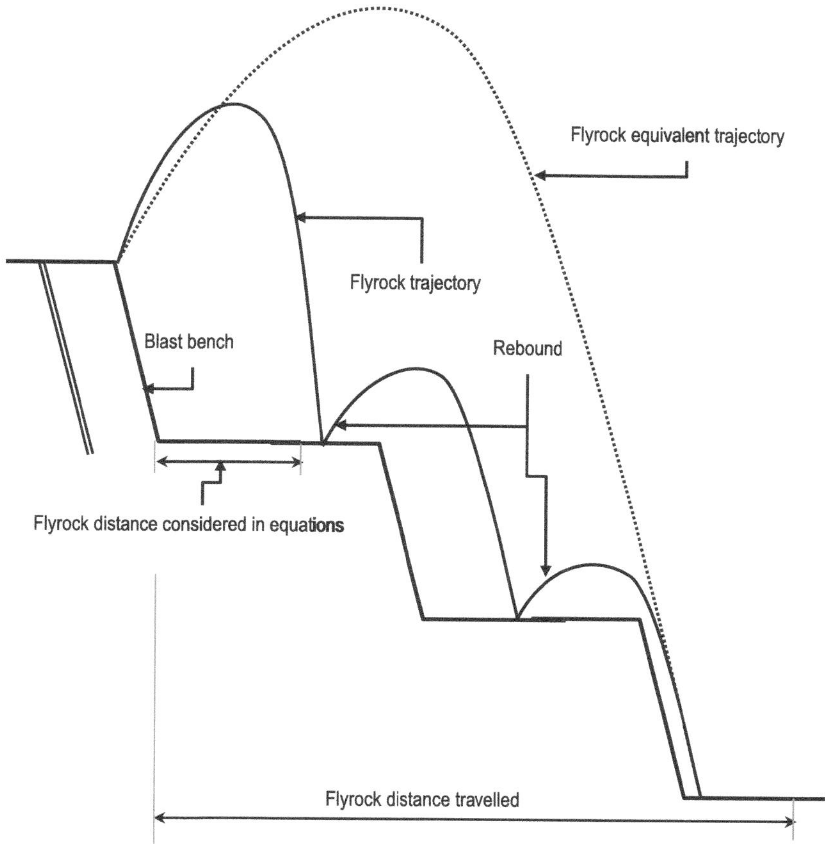

FIGURE 2.4 The concept of multiple rebounds in bench formation.

to a significant distance and hence the difference in calculations. It may be mentioned that the results provided here are from controlled experiments with sufficient available distance in front and back of the blast that created ideal conditions for the rebound and the situations can vary significantly in full-scale production blasts.

The objective of providing the details over here is that the phenomenon of the rebound has been completely ignored in analysis and researchers can think relooking into their data and analysis and future attempts can account for this important aspect of flyrock. In addition, the aforementioned observations raise some pertinent questions about the flyrock distance calculations.

There are two possibilities that might have been considered in the prediction flyrock distance:

1. The total distance of travel has been measured and this distance of travel of flyrock is included in the flyrock distance prediction equations.

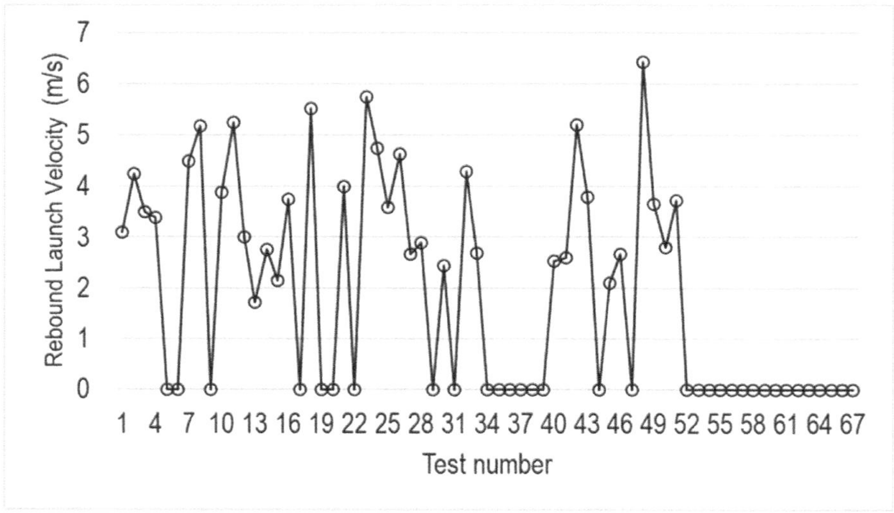

FIGURE 2.5 Initial velocity of the rebound of the flyrock.

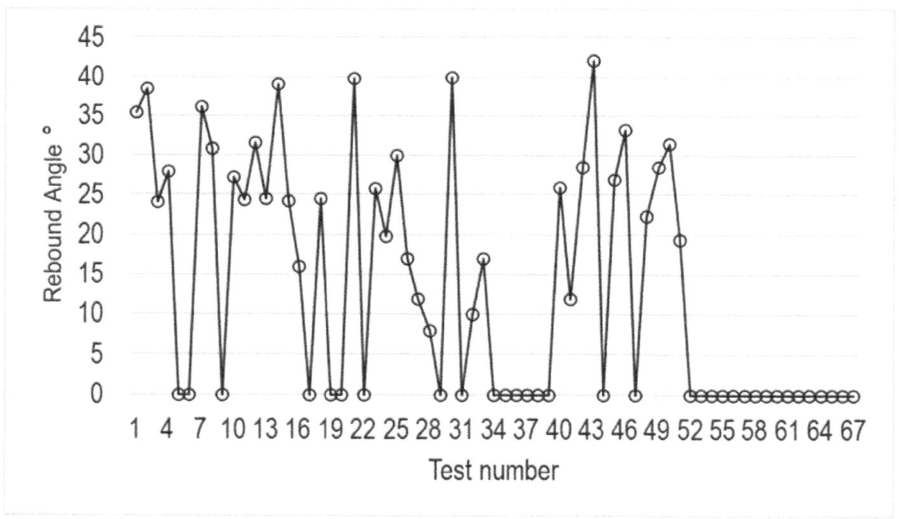

FIGURE 2.6 Angle of rebound of the flyrock.

2. The predicted distance of travel of flyrock based on ballistic equations has been used. This condition, if invoked, does not consider rebound and hence a correction to such equations will be inevitable. In such cases, the equivalent trajectory may be used.

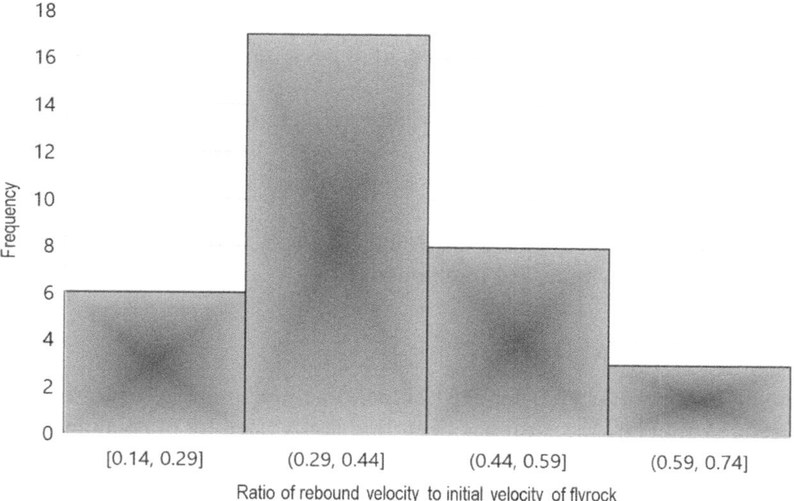

FIGURE 2.7 Frequency of ratio of initial velocity to the velocity after rebound of flyrock.

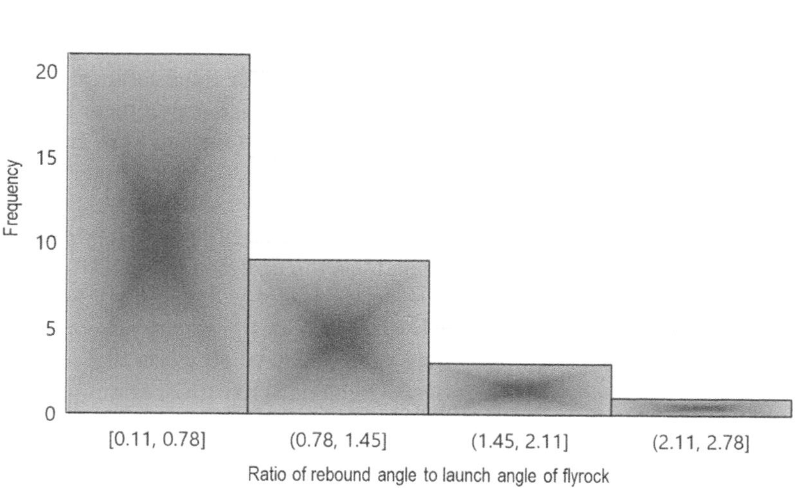

FIGURE 2.8 Frequency of ratio of rebound angle to the initial launch angle of flyrock.

The conditions that will stop the rebound of the flyrock from the surface are as follows:

1. Presence of high drag and frictional surfaces on which the flyrock lands.
2. Presence of muddy, sand, or wet soil landing place that will not allow the rock to reflect from the surface.

3. Presence of a barrier about the landing place of flyrock that will restrict further movement of the flyrock and arrest the same.
4. Landing of the flyrock on roof or wall of a house/structure and penetrating the same.
5. Landing of the flyrock within structures where it has no further space to reflect.
6. Deep pit conditions which do not allow flyrock to rebound.

However, if the landing surface is hard, there is every chance that the flyrock will travel further under the influence of rebound.

2.4 BOULDER BLASTING

Although Wentworth (1922) defined boulders as round particles of 0.259–4.096 m for clastic sediments, the term has been frequently used by blasting engineers to define large fragments. In such a case, the definition of boulder in mining and excavations may vary, as it is dependent on the size of the bucket of a shovel. Generally, the best fragment size for a shovel bucket is 0.15–0.2 times the nominal diameter of the bucket, i.e. the cube root of the volume of the bucket (Rzhevskiĭ, 1985). So, the definition of boulder for a blast can vary from one mine to other. Boulder may at the best be described as oversize rock fragment for a particular mining condition. However, for the sake of simplicity, "boulder" is preferred for use over here.

Such oversize rock fragments are frequently encountered in mining and other blasting operations like roads, railways, slopes, etc. The slope failures and fall of boulders on railway tracks and roads are common problems, and these hamper the movement of trains and vehicles (Bhagat et al., 2021). One can even encounter in situ boulders in soil formations while excavating foundations of bridges, houses, and other infrastructure. In mining operations also, one can find large-size boulders produced during bench blasting that may be due to poor blast design or presence of boulders in loose strata. Since handling and loading of oversize boulders is difficult or not possible, the same need to be further fragmented. Hydraulic rock breakers and different such techniques (Murray et al., 1994), designated as secondary breaking, are frequently deployed to fragment such boulders. However, the operation is costly owing to mechanical methods of rock breaking (Kristin & Maras, 2018). This is the reason that in many such situations, people resort to use of explosives to break such boulders.

Boulder blasting is a special application of explosive use and is different from bench blasting, as the number of free faces available is multiple. From the perspective of flyrock, the blasted fragments can be projected in all directions and hence pose a severe risk to the objects of concern.

There are two main methods that are deployed in breaking the boulders:

1. **Plaster-shooting**, also known as mud capping (Konya & Walter, 1991): In this method, the explosive in cartridge or other form is just placed on the oversize rock or even covered with mud and exploded (Figure 2.9).

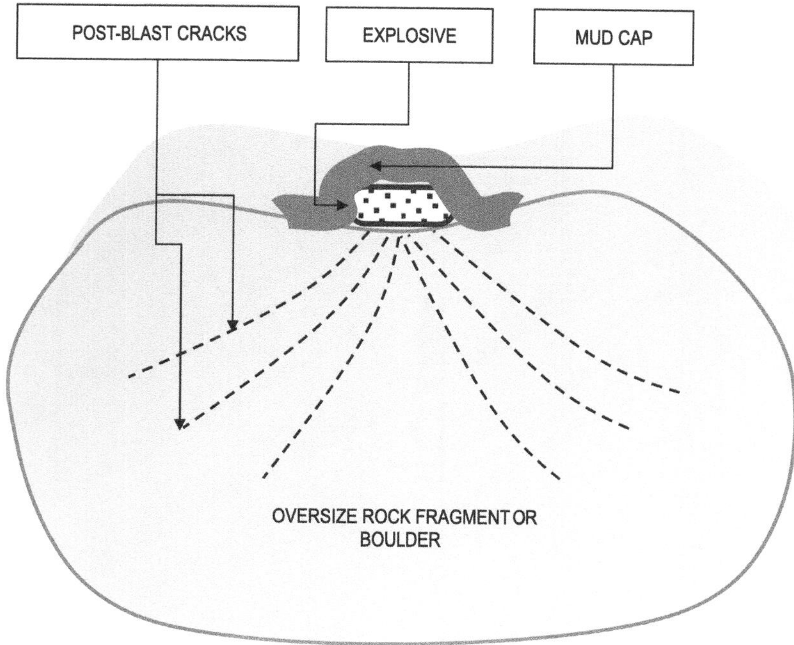

FIGURE 2.9 The process of plaster shooting explained.

This generally fragments the rock, if it is not too hard, and the explosive quantity is less for the size of the boulder being blasted. However, there are least possibilities of flyrock in such cases as gases escape directly into the open and dissipate fast due to lack of confinement.

2. **Pop-shooting**, also known as block holing or boulder busting (Konya & Walter, 1991): In this method, a small-diameter hole is drilled in the oversize rock and explosive is placed within the hole and detonated (Figure 2.10), hence fragmenting the rock block into smaller pieces. However, the process presents an ideal case for flyrock that can shoot several hundred metres (Bhagat et al., 2021), depending upon the quantity of the explosive used for blasting. Other attributes of such flyrock are like what is being discussed in this work. Little literature is available on flyrock emanating from boulder blasting. In one such study (Bhagat et al., 2021), 61 boulders from a site of Konkan Railways (India) were monitored while being blasted. The data of flyrock was analysed using multiple linear regression (MLR) and classification and regression trees (CART) techniques.

It is evident that pop-shooting presents a typical case of least confinement with potential to generate flyrock.

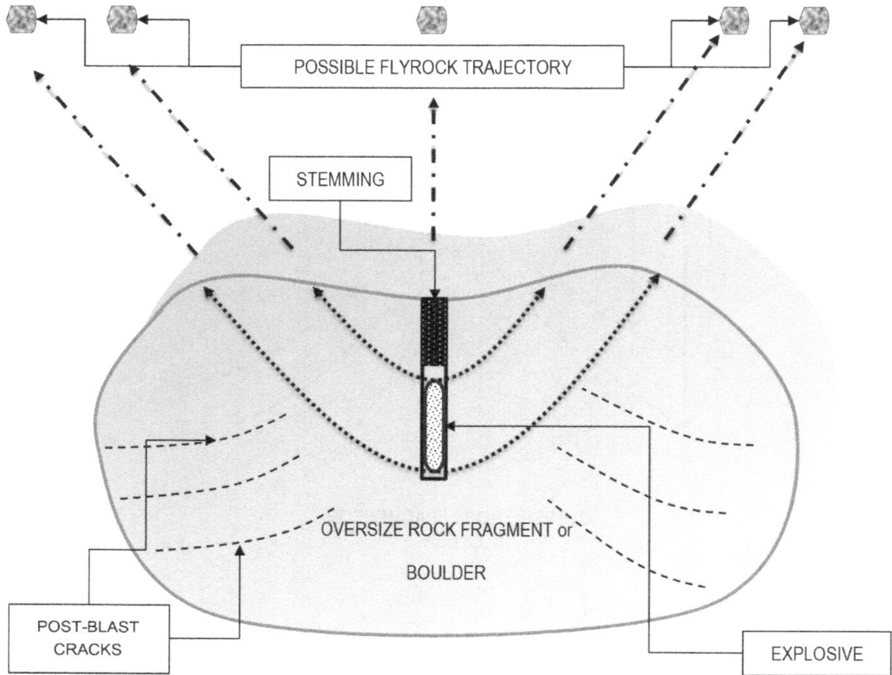

FIGURE 2.10 The process of pop-shooting explained—this presents an ideal case for generation of flyrock.

2.5 TOE BLASTING

Maintaining a proper level of benches is important during blasting. In hard rock conditions, there is every possibility that the toe of the bench is not broken during blasting. If unattended, the toe can result in a gradient of bench towards the bench and shorten its height over time. This is the main purpose of sub-drilling wherein the blasthole is drilled more than the bench height to break the toe. However, under certain circumstances, the toe does not break and requires secondary blasting in which smaller depth blastholes are drilled in an inclined manner in the toe and blasted. Since the toe is uneven, blastholes are confined and angular, the chances of flyrock are quite high.

2.6 SITUATIONS OF FLYROCK

Flyrock is encountered in many situations, and these differ from one another in terms of the scale of the operation, the frequency of blasts, the quantity of explosive used, nearness of localities, and several other conditions. These can be classified as mining, civil and construction, large chambers in underground, slope, and demolition blasting that are briefly introduced here considering flyrock occurrence.

2.6.1 Mining

Mining presents the major concern in the case of flyrock as the industry consumes most of the explosives and accordingly involves very high frequency of blasting, thus presenting greater probabilities of flyrock. In the case of underground mining, two situations arise: one is purely related to sort of tunnelling for development and the other is related to massive blasting and development of stopes in metal mines. Unless a huge chamber is developed for storage or other purposes, blasting in tunnelling has limitations of a single face, i.e. it is highly confined. The throw of the broken rock is always towards the open area and hence possibilities of impact by a flyrock arise. However, since such flyrock(s) are restricted in space by the walls of the tunnel, there are practically no chances of it hitting objects, unless these have been left negligently or deliberately at the blast site. The case of large chambers is of interest, as most of such excavations are carried out with normal bench blasting method and, there are possibilities of flyrock hitting people and infrastructure within the chamber. Nonetheless, the range of flyrock is restricted to the dimensions of the chamber. The rules of flyrock, as applicable to surface blasting, in such cases will be similar.

2.6.2 Civil and Construction

Nowadays civil and other such constructions are expanding and present a case for flyrock as such blasting are conducted in vicinity of the habitats, busy roads, and other infrastructures. Blasts conducted in urban environment have high possibilities of damage to structures and injuries to personnel due to flyrock (Mohamad et al., 2018; Venkatesh et al., 2013). Such sites are particularly more vulnerable to flyrock occurrence due to space limitations and occurrence of habitats or public utilities nearby. One major issue with such blasts is that they are mostly unregulated as these do not fall under domain of mining.

2.6.3 Other Underground Excavations

The blasts in underground do not present any threat from the flyrock, except for damage to the equipment and other installations underground. The only case which needs consideration is excavation of large chambers for hosting of underground facilities and defence infrastructure. In such cases, blasts are just equivalent to surface blasts. However, the flyrock, if generated, will be restricted to the exposed area and dimensions of the excavation. There are practically no examples of accidents in such case due to flyrock.

2.6.4 Slopes

The blasting in slopes falls in the domain of civil construction. Slopes are generally intersected by road or railway lines for which blasting is conducted. In such cases, the possibilities of flyrock are higher and even if restricted, any such fragment can travel to greater distance if the same rolls down the slopes.

2.6.5 Demolition Blasting

Demolition blasting is a special control blasting technique to bring down high-rise structures, large buildings, bridges, chimneys, and even breaking debris of collapsed buildings. The method, along with its benefits, has been discussed in detail in several texts. The method is believed to have originated during World War II for reconstruction of buildings damaged during the war (Jimeno et al., 1995). The method has evolved significantly since then. The technique of demolition blasting is summarized by Sinitsyn (2018). Away from breaking concrete, there are instances where the method has been used even to break steel into smaller parts for easy transportation (Fan et al., 2002; Kato et al., 1998) and other structures like bridges.

Demolition blasting has every probability of generating flyrock as the construction elements of buildings or structures are small in nature and can break in any direction. The method can at the best be defined as a modified pop-shooting technique, which has been demonstrated earlier with possibilities of flyrock. This is one of the reasons that such technique involves huge costs and risk-sharing through insurance. However, there are lot of examples in the public domain where there have been no instances of flyrock owing to advancement of technique and invariable provisions for covering of the demolition area. Since small-diameter blastholes are deployed, these result in small fragments that do not travel far and can be trapped by simple muffling of the buildings.

2.7 DOMAIN SHIFT TO ROCKFALL

A special case of rolling rock fragments from slopes has recently been referred to as flyrock. There are host of such reports, analysis, risk assessment, and management measures presented by several authors, e.g. Briones-Bitar et al. (2020), and is beyond scope of this work. A comprehensive treatment of the subject is provided by Prades-Valls et al. (2022). Inclusion of flyrock or "fly rock" in rockfall studies is practically confusing to the readers as there is no role of explosives in such failures and bouncing rocks. Unless rocks or boulders are thrown from a blast being conducted on the hills (Collins et al., 2022), the case of rolling rocks may be assigned a different class or name. The roll down or bouncing of fragments from blasting at higher benches may be the only common concern.

2.8 FLYROCK MEASUREMENT METHODS

Unlike ground vibrations, flyrock does not have a standard method of prediction. Flyrock must be identified, its distance calculated or practically measured, and size determined. This needs proper instrumentation and methodology. Since such measurements are used in estimating initial velocity and launch angle of flyrock, it is pertinent to have a look at the methods available currently.

The occurrence of flyrock can be observed through cameras that record at a maximum of 30 frames per second (fps). High-speed cameras which have capacity to record motion at several thousand frames in a second are used in flyrock studies for estimation of its initial velocity, launch angle, and identification of its trajectory.

Blair (1960) provided a comprehensive detail of deployment, correction, and analysis of videos from high-speed cameras along with some earliest citations (not traceable online) of such use. The historical instance of use of high-speed photography by Giltner and Worsey (1986) documented recorded events of cast blasting along with the interpretation of their results.

The use of comprehensive instrumentation for blast recording along with the high-speed video monitoring is recommended by Chiappetta (1998) to capture complete data on blasting. The benefits of use of such cameras have also been documented by several other authors, e.g. Adermann (2007), Adermann et al. (2015), Adkins (1986), Gong and Wu (2019), and Qian (1994). Recent development of high-resolution, high-speed cameras have made it possible to measure several aspects of blasting, e.g. blast design (Chiappetta et al., 1983), rock fracture process (Bamford et al., 2020; Takahashi et al., 2018; Tang & Ding, 2019; Yang et al., 2018), delay performance (Winzer et al., 1979), etc. Readers interested in the topic are encouraged to go through the paper of Adermann et al. (2015) as they provide details of high-speed cameras along with cost details and other pros and cons of such systems. There are several concerns while measuring flyrock velocities with the help of high-speed videos:

Type of Camera: The modern-day cameras come with good resolution and there are lot of manufactures that supply such cameras. A trade-off will be needed to decide the type and number of cameras required for capturing the blast event. Fortunately, there are some standard manufacturers who provide cameras with excellent resolution of up to 1000 frames/s that are inexpensive. Users can opt for several such cameras to have a full view of the blast and flyrock, if any.

Camera Placement: The placement of camera should be in such a position that is safe, oriented across the face, and preferably on the same bench on which the blast is being conducted. One more consideration is that the initiation should be towards the camera position so that the movement of the fragments throughout the blast is captured by the camera. If the firing sequence of blasthole is away from the camera position, the movement of fragments will be obliterated by the rockmass dislodged first in the sequence. If a high-speed camera is to be deployed, there should be high degree of object illumination.

Focal Length and Camera Zoom: Patterson (1957) recommends that 25 is the best focal length for a camera. Optical zoom of 40–75× should be sufficient to have provide excellent images and information of the flyrock. However, one can miss the flyrock movement if zoom is too high and the fragment information or tracking of wide angle is used. A simple rule is "zoom in and record the breakage, zoom out and see flyrock but it will be difficult or impossible to track the motion owing to loss of fragment in the image or change of reference frame." Accordingly, a user should decide in advance about what will be the best zoom to capture the flyrock fragment in frames and trace it to a significant distance.

Frame Rates: Frame rates will vary for specific objectives of video monitoring. Tracking a delay element, fracturing in rockmass, etc. requires frame rates of >1000 fps. It is not always necessary to monitor a blast at very high frame rates. Feng et al. (2011) concluded that if a particle is undergoing acceleration, the error at high frame rate increases due to uncertain position of the fragment. They suggested that the record should be made at higher frame rates and particle tracking should be done by skipping intermediate frames while looking for best candidates. This will allow tracking of fragments and logging these for analysis of velocity without missing the details.

Ground Shaking: Since vibration travels quite fast, it induces shaking in the camera placed on a tripod or stand. This can result in defocused or blurred images and errors due to movement of frame of reference that induces errors in measurement. There are software, e.g. VirtualDub,[3] that use DeShaker[4] movie stabilizer to remove the shaking effect and can be deployed but require considerable effort to vectorize the images.

Motion Blur: Due to high speed of the fragments moving, it is possible to record images with motion blur. Some cameras have in-built features for correcting motion blur and should be checked before procuring the same.

Scaling: Placement of markers on the ground around the blast is essential to make proper measurements of the movement of rock fragments or blasted muck. The markers should be of sufficient height and width and preferably bright coloured having excellent visibility. Since the measurements are in two directions, multiple cameras may be required to have a complete motion analysis.

Motion Analysis: Some expensive high-speed cameras are supplied with motion analysis software that allow tracking of rock fragments in motion and provide automatic calculations of the several parameters, including velocity in two directions, the angle, and derivatives like acceleration. Tracker,[5] an open-source software, provides several advanced features to conduct a quantifiable motion analysis. It is expected that the biomechanical movement systems (Payton & Bartlett, 2017) will advance in near future for flyrock measurements also.

2.8.1 Use of Drones in Flyrock Videography

Drones have established their utility in blasting in present scenario. The applications of drones in mining have been reviewed by Bhatawdekar et al. (2018). Drones can be quite useful in determining the kinematics of flyrock, e.g. Prades-Valls et al. (2022) captured videos with drones and developed a 3D system to measure different aspects of the rock fall kinematics and fragmentation. Drones have an advantage that these can be controlled remotely, fixed at any place in air near the blast, and capture videos without shaking. Since drones are GPS enabled, the errors in analysis can be corrected with a fair degree of accuracy owing to the advantage of known location, i.e. both the blast site and the final landing position of the flyrock. However, the frame rate of drone-mounted cameras is mostly 30, which is quite less for capturing flyrock for proper analysis. The use of high-speed cameras on drones may be expected in near future.

NOTES

1. www.mainelegislature.org/legis/statutes/38/title38sec490-W.html (20.04.2022)
2. https://en.wikipedia.org/wiki/Flyrock (20.05.2022)
3. www.virtualdub.org/index.html (11.09.2022)
4. www.guthspot.se/video/deshaker.htm (11.09.2022)
5. https://physlets.org/tracker/ (11.09.2022)

REFERENCES

Adermann, D. (2007). What information and production benefit can i gain by using digital high-speed camera and data capture technology? *Australasian Institute of Mining and Metallurgy Publication Series*, 187–188.

Adermann, D., Chalmers, D., Martin, C., & Wellink, S. (2015). High-speed video: An essential blasting tool. *AusIMM Bulletin, 2015*, 28–32.

Adkins, N. (1986). Aids to blast design. *Quarry Management, 13*(10), 35–36, 38.

Anh Tuan, N., van Hoa, P., van Viet, P., Dinh Bao, T., & Thi Hai, L. (2020). Simulation of fly-rock distance as a function of blast conditions: A case study in Vietnam. *Inzynieria Mineralna, Journal of the Polish Mineral Engineering Society, 2020*. https://doi.org/10.29227/IM-2020-02-33

Bamford, T., Medinac, F., & Esmaeili, K. (2020). Continuous monitoring and improvement of the blasting process in open pit mines using unmanned aerial vehicle techniques. *Remote Sensing, 12*(17), 2801. https://doi.org/10.3390/rs12172801

Benwei, L., Dezhi, C., Yingjun, Z., & Qi, W. (2016, August). The safety evaluation of blasting fly-rock base on unascertained measurement theory. In *International conference on smart grid and electrical automation (ICSGEA)* (pp. 344–346). IEEE.

Bhagat, N. K., Rana, A., Mishra, A. K., Singh, M. M., Singh, A., & Singh, P. K. (2021). Prediction of fly-rock during boulder blasting on infrastructure slopes using CART technique. *Geomatics, Natural Hazards and Risk, 12*(1), 1715–1740. https://doi.org/10.1080/19475705.2021.1944917

Bhatawdekar, R. M., Choudhury, S., & Modmad, E. T. (2018). Uav applications on projects monitoring in mining and civil engineering. *Journal of Mines, Metals and Fuels, 66*(12), 867–872.

Blair, B. E. (1960). *Use of high-speed camera in blasting studies* (Vol. 5584). US Department of the Interior, Bureau of Mines.

Briones-Bitar, J., Carrión-Mero, P., Montalván-Burbano, N., & Morante-Carballo, F. (2020). Rockfall research: A bibliometric analysis and future trends. *Geosciences (Switzerland), 10*(10), 1–25. https://doi.org/10.3390/geosciences10100403

Chiappetta, R. F. (1998). Blast monitoring instrumentation and analysis techniques, with an emphasis on field applications. *Fragblast, 2*(1), 79–122. https://doi.org/10.1080/13855149809408880

Chiappetta, R. F., Bauer, A., Dailey, P. J., & Burchell, S. L. (1983). Use of high-speed motion picture photography in blast evaluation and design. In *Proceedings of the annual conference on explosives and blasting technique* (pp. 258–309). International Society of Explosive Engineers.

Choudhary, B. S., & Rai, P. (2013). Stemming plug and its effect on fragmentation and muckpile shape parameters. *International Journal of Mining and Mineral Engineering, 4*(4), 296–311. https://doi.org/10.1504/IJMME.2013.056854

Collins, B. D., Corbett, S. C., Horton, E. J., & Gallegos, A. J. (2022). Rockfall kinematics from massive rock cliffs: Outlier boulders and flyrock from Whitney Portal, California, rockfalls. *Environmental and Engineering Geoscience, 28*(1), 3–24. https://doi.org/10.2113/EEG-D-21-00023

Davies, P. A. (1995). Risk-based approach to setting of flyrock "danger zones" for blast sites. *Transactions—Institution of Mining & Metallurgy, Section A*, *104*, 96–100. https://doi.org/10.1016/0148-9062(95)99212-g

Dhekne, P. Y. (2015). Environmental impacts of rock blasting and their mitigation. *International Journal of Chemical, Environmental & Biological Sciences (IJCEBS)*, *3*(1), 46–50.

Dick, R. A., Fletcher, L. R., & D'Andrea, D. V. (1983). Explosives and blasting procedures manual. *Information Circular—United States, Bureau of Mines*, *8925*.

Fan, L., Shen, W., & Li, Y. (2002). The causes of flyrock and safety precautions in demolition blasting. *Engineering Blasting of China*, *8*(1), 35–38. www.matec-conferences.org/articles/matecconf/pdf/2018/29/matecconf_spbwosce2018_03019.pdf

Feng, Y., Goree, J., & Liu, B. (2011). Errors in particle tracking velocimetry with high-speed cameras. *Review of Scientific Instruments*, *82*(5), 053707. https://doi.org/10.1063/1.3589267

Fletcher, L. R., & D'Andrea, D. V. (1990). Control of flyrock in blasting. *Journal of Explosives Engineering*, *7*(6), 167–177.

Frasher, B., & Dambov, R. (2018). Fly rocks in surface mine during the blasting. In *XIth profession advice on topic: Technology of underground and surface exploitation of mineral raw materials PODEX—MORE'18* (pp. 113–119). http://eprints.ugd.edu.mk/20917/1/014. Frasher—Dambov—Fly rocks during blasting.pdf

Gibson, M. F. L., & St George, J. D. (2001). Implications of flyrock associated with blasting in urban areas. *AusIMM NZ Branch Conference*, 10p.

Giltner, S. G., & Worsey, P. N. (1986). Blast monitoring using high speed video research equipment. In *Proceedings of the twelfth conference on explosives and blasting technique*. International Society of Explosives Engineers, 6p.

Gong, M., & Wu, H. (2019). High-speed photography image acquisition system in tunnel blasting and parameters study on precisely controlled blasting. *Baozha Yu Chongji/Explosion and Shock Waves*, *39*(5). https://doi.org/10.11883/bzycj-2018-0319

Guo, H., Nguyen, H., Bui, X. N., & Armaghani, D. J. (2019). A new technique to predict fly-rock in bench blasting based on an ensemble of support vector regression and GLMNET. *Engineering with Computers*, 1–15.

Guo, H., Nguyen, H., Bui, X. N., & Armaghani, D. J. (2021). A new technique to predict fly-rock in bench blasting based on an ensemble of support vector regression and GLMNET. *Engineering with Computers*, *37*(1), 421–435. https://doi.org/10.1007/s00366-019-00833-x

Han, H., Armaghani, D. J., Tarinejad, R., Zhou, J., & Tahir, M. M. (2020). Random forest and bayesian network techniques for probabilistic prediction of flyrock induced by blasting in quarry sites. *Natural Resources Research*, *29*(2), 655–667. https://doi.org/10.1007/s11053-019-09611-4

Hosseini, S., Poormirzaee, R., Hajihassani, M., & Kalatehjari, R. (2022). An ANN-fuzzy cognitive map-based Z-number theory to predict flyrock induced by blasting in open-pit mines. *Rock Mechanics and Rock Engineering*, 1–18. https://doi.org/10.1007/s00603-022-02866-z

Huang, X., Zhang, X., Sun, J., Hao, H., & Bai, R. (2020). Research on and application of blasting excavation technology for highway cutting under complex environment. *IOP Conference Series: Earth and Environmental Science*, *571*(1). https://doi.org/10.1088/1755-1315/571/1/012094

IME. (1997). *Glossary of commercial explosives industry terms*. Institute of Makers of Explosives.

Jimeno, C. L., Jimeno, E. L., & Carcedo, F. J. A. (1995). *Drilling and blasting of rocks*. Balkema. https://doi.org/10.1201/9781315141435-20

Kane, W. F. (1997). Drilling and blasting of rocks. *Environmental & Engineering Geoscience, III*(1), 154. https://doi.org/10.2113/gseegeosci.iii.1.154

Kato, M., Nakamura, Y., Matsuo, A., Ogata, Y., Katsuyama, K., & Hashizume, K. (1998). Study on the impact failure of steel plates and effective shape of shaped charges used for blasting demolition of steel multi-story buildings. *Kayaku Gakkaishi/Journal of the Japan Explosives Society*, *59*(5).

Konya, C. J., & Walter, E. J. (1991). *Rock blasting and overbreak control* (No. FHWA-HI-92–001; NHI-13211). National Highway Institute, 430p.

Kopp, J. W. (1994). Observation of flyrock at several mines and quarries. In *Proceedings of the conference on explosives and blasting technique* (pp. 75–81). International Society of Explosive Engineers.

Kristin, S., & Maras, M. (1994). Secondary rock breaking by use of impactors. In *Geomechanics 93* (pp. 441–444). Routledge.

Little, T. N. (2007). Flyrock risk. *Australasian Institute of Mining and Metallurgy Publication Series*, 35–43.

Lwin, M. M., & Aung, Z. M. (2019). Prediction and controlling of flyrock due to blasting for Kyaukpahto gold mine. *International Journal of Advances in Scientific Research and Engineering*, *5*(10), 338–346. https://doi.org/10.31695/ijasre.2019.33574

Mohamad, E. T., Yi, C. S., Murlidhar, B. R., & Saad, R. (2018). Effect of geological structure on flyrock prediction in construction blasting. *Geotechnical and Geological Engineering*, *36*(4), 2217–2235. https://doi.org/10.1007/s10706-018-0457-3

Moore, A. J., & Richards, A. B. (2011). Blasting near mine surface infrastructure. In *EXPLO 2011—blasting—controlled productivity* (pp. 201–209). AusIMM.

Murray, C., Courtley, S., & Howlett, P. F. (1994). Developments in rock-breaking techniques. *Tunnelling and Underground Space Technology Incorporating Trenchless*, *9*(2). https://doi.org/10.1016/0886-7798(94)90034-5

Patterson, E. M. (1957). Photography applied to the study of rock blasting. *The Journal of Photographic Science*, *5*(6). https://doi.org/10.1080/00223638.1957.11736612

Payton, C. J., & Bartlett, R. M. (2017). *Biomechanical evaluation of movement in sport and exercise: The British association of sport and exercise sciences guidelines*. Routledge. https://doi.org/10.4324/9780203935750

Persson, P. A., Holmberg, R., & Lee, J. (1994). *Rock blasting and explosives engineering*. CRC Press. https://doi.org/10.5860/choice.31-5469

Prades-Valls, A., Corominas, J., Lantada, N., Matas, G., & Núñez-Andrés, M. A. (2022). Capturing rockfall kinematic and fragmentation parameters using high-speed camera system. *Engineering Geology*, *302*, 106629. https://doi.org/10.1016/j.enggeo.2022.106629

Qian, L. (1994). The application of high speed video system for blasting research. In *Proceeding of the conference on explosives and blasting technique* (pp. 359–375). International Society of Explosive Engineers.

Qiao, W. (2020). Application of controlled blasting technology in mountain excavation under complex conditions. *IOP Conference Series: Earth and Environmental Science*, *558*(2). https://doi.org/10.1088/1755-1315/558/2/022045

Raina, A. K., Chakraborty, A. K., Ramulu, M., & Choudhury, P. B. (2006). *Flyrock prediction and control in opencast metal mines in India for safe deep-hole blasting near habitats—a futuristic approach* (Report). CSIR-Central Institute of Mining and Fuel Research, 156p.

Raina, A. K., & Murthy, V. M. S. R. (2016). Importance and sensitivity of variables defining throw and flyrock in surface blasting by artificial neural network method. *Current Science*, *111*(9), 1524–1531. https://doi.org/10.18520/cs/v111/i9/1524-1531

Raina, A. K., Murthy, V. M. S. R., & Soni, A. K. (2015). Flyrock in surface mine blasting: Understanding the basics to develop a predictive regime. *Current Science*, *108*(4), 660–665. https://doi.org/10.18520/CS/V108/I4/660-665

Raina, A. K., Soni, A. K., & Murthy, V. M. S. R. (2013). Spatial distribution of flyrock using EDA: An insight from concrete model tests. In *Rock fragmentation by blasting, proceedings of the 10th international symposium on rock fragmentation by blasting, Fragblast 10* (pp. 563–568). Taylor & Francis Books Ltd. https://doi.org/10.1201/b13759-79

Roth, J. A. (1979). *A model for the determination of flyrock range as a function of shot conditions* (Report No. PB81222358). US Department of Commerce, NTIS, 61p.

Rustan, A., Cunningham, C. V. B., Fourney, W., & Spathis, A. (2011). Mining and rock construction technology desk reference. In A. Rustan (Ed.), *Mining and rock construction technology desk reference*. CRC Press. https://doi.org/10.1201/b10543

Rzhevskiĭ, V. V. (1985). *Opencast mining: Unit operations*. Mir Publishers.

Sawmliana, C., Hembram, P., Singh, R. K., Banerjee, S., Singh, P. K., & Roy, P. P. (2020). An investigation to assess the cause of accident due to flyrock in an opencast coal mine: A case study. *Journal of The Institution of Engineers (India): Series D*, *101*(1), 15–26. https://doi.org/10.1007/s40033-020-00215-4

Sharma, A., Mishra, A. K., & Choudhary, B. S. (2019). Impact of blast design parameters on blasted muckpile profile in building stone quarries. *Annales de Chimie: Science Des Materiaux*, *43*(1). https://doi.org/10.18280/acsm.430105

Shevkun, E. B., Leschinsky, A. V., & Piotrovich, A. A. (2018). Without fly rock blasting technology of railway reconstruction. *IOP Conference Series: Materials Science and Engineering*, *463*(2). https://doi.org/10.1088/1757-899X/463/2/022081

Sinitsyn, D. (2018). Drilling-and-blasting method of demolition. *MATEC Web of Conferences*, *170*, 3019.

Siskind, D. E., & Kopp, J. W. (1995). Blasting accidents in mines: A 16-year summary. In *21st annual conference on explosives and blasting technique* (pp. 224–239). International Society of Explosives Engineers.

Stojadinović, S., Pantović, R., & Žikić, M. (2011). Prediction of flyrock trajectories for forensic applications using ballistic flight equations. *International Journal of Rock Mechanics and Mining Sciences*, *48*(7), 1086–1094. https://doi.org/10.1016/j.ijrmms.2011.07.004

Takahashi, Y., Saburi, T., Sasaoka, T., Wahyudi, S., Kubota, S., Shimada, H., & Ogata, Y. (2018). Fundamental study on rock fracture mechanism induced by blasting in small-scale blasting tests. *Science and Technology of Energetic Materials*, *79*(5–6), 175–179.

Tang, M. W., & Ding, Y. C. (2019). High-speed image analysis of the rock fracture process under the impact of blasting. *Journal of Mining Science*, *55*(4). https://doi.org/10.1134/S1062739119045928

Tao, T., Huang, P., Wang, S., & Luo, Y. (2018). Safety evaluation of blasting fly-rock based on unascertained measurement model. *Instrumentation Mesure Metrologie*, *17*(1), 55–62. https://doi.org/10.3166/I2M.17.55-62

Venkatesh, H. S., Gopinath, G., Balachander, R., Theresraj, A. I., & Vamshidhar, K. (2013). Controlled blasting for a metro rail project in an urban environment. In *Rock fragmentation by blasting, proceedings of the 10th international symposium on rock fragmentation by blasting, FRAGBLAST 10* (pp. 793–801). CRC Press/Balkema. https://doi.org/10.1201/b13759-114

Wentworth, C. K. (1922). A scale of grade and class terms for clastic sediments. *The Journal of Geology*, *30*(5), 377–392. https://doi.org/10.1086/622910

Winzer, S. R., Furth, W., & Ritter, A. (1979). Initiator firing times and their relationship to blasting performance. In *20th U.S. symposium on rock mechanics, USRMS 1979* (pp. 461–470). American Rock Mechanics Association.

Workman, J. L., & Calder, P. N. (1994). Flyrock prediction and control in surface mine blasting. *Proceedings of the conference on explosives and blasting technique* (pp. 59–74). International Society of Explosives Engineers.

Xiao, S., Ding, X., Ma, L., & Dong, G. (2019). Parameters optimization of dragline working platform based on nonlinear programming. *Meitan Xuebao/Journal of the China Coal Society, 44*(10). https://doi.org/10.13225/j.cnki.jccs.2018.1462

Yang, L., Ding, C., Yang, R., Lei, Z., & Wang, J. (2018). Full field strain analysis of blasting under high stress condition based on digital image correlation method. *Shock and Vibration, 2018*, 1–7. https://doi.org/10.1155/2018/4894078

Yao, J., Wang, G., Jin, Z., & Shun, Y. (2011). Blasting and controll technique of danger rock mass at Wuzhaoguan on Yichang-Wanzhou Railway. In *Proceedings of the 2nd international conference on mechanic automation and control engineering, MACE 2011* (pp. 2971–2973). IEEE. https://doi.org/10.1109/MACE.2011.5987612

Zhou, J., Aghili, N., Ghaleini, E. N., Bui, D. T., Tahir, M. M., & Koopialipoor, M. (2020). A Monte Carlo simulation approach for effective assessment of flyrock based on intelligent system of neural network. *Engineering with Computers, 36*(2), 713–723. https://doi.org/10.1007/s00366-019-00726-z

Zijun, W. (2002). Explanation and discussion on flying stones accident in blasting operation. In *The 7th international symposium on rock fragmentation by blasting* (pp. 672–675). Metallurgical Industry Press.

3 The True Nature of Flyrock

The objective of a blast, as detailed in Chapter 1, is to fragment the rockmass into optimum size, so that it can be loaded, hauled, and processed further in an economically efficient manner. Mere fragmenting the rockmass requires explosive quantity that is not sufficient to displace the rock. In order to throw and heave the rock for proper loading, extra amount of explosive is required to move the centre of gravity of the broken rock by a few bench heights (Langefors & Kihlstrom, 1978). Accordingly, the throw of the broken rock or muck will increase with increase in explosive weight used in a blasthole. In certain conditions, some broken fragments launch independent of the heaving process of the muck, travel farther than designed, and transform into flyrock. It is thus essential to inquire and identify such special conditions leading to generation of flyrock and to define its true nature.

In view of the recent developments, particularly related to use of artificial intelligence techniques (included in Chapter 4), it is essential to have an insight into the role of blast design and rockmass in flyrock generation to put its occurrence into perspective. This will help to build a firm basis for critical examination of the predictive models proposed by various authors.

An attempt has been made here to document all known modes of flyrock that may eject out of a blast face owing to several reasons. These include blast design, drill diameter, stemming length and type, and defects in rockmass like fractured faces and stemming zones and weak zones etc. along with other reasons. It is amply clear that the mechanisms of flyrock are varied in nature and yet are not regular in their occurrence. Thus, a distinction is made in an engineered blast and a blast with unknowns or ignored situations that can result in flyrock. It is important to note that the conditions of unknowns can qualify as errors. However, if such conditions are known and yet not considered during blasting, it can be classified as a mistake. Hence, any flyrock distance predictive mechanisms should take these conditions into account.

3.1 BLAST DESIGN AND FLYROCK

The blast design variables, as introduced earlier (Chapter 1), include blasthole depth, burden, spacing, and stemming (length). Other quantities that are indirectly involved in blast design are total explosive per hole, total explosive per blast, quality of explosive, length of explosive with respect to blasthole length, and specific charge. There are instances that may or may not be corroborated, where blast design has been directly assigned as the reason for occurrence of flyrock, irrespective of causative factors. A comprehensive list of such reports will be provided in Chapter 4. However, a question arises—whether a perfectly engineered blast should generate a flyrock or not? An engineering judgement of the query should be negative and

DOI: 10.1201/9781003327653-3

scientifically, a perfectly engineered blast should not generate flyrock for obvious reasons. Pal Roy (2005) indicated that all blasts can generate flyrock with higher distances in poorly controlled blasts. The qualifier "poorly controlled" in the case of flyrock has a distinct connotation as "engineered" and "poorly controlled" blasts contradict each other.

The only condition that favours the flyrock occurrence, in an engineered blast, is loading of a blasthole with excess explosive than required, to fragment and heave the broken rockmass. Nonetheless, blasting in itself is not a perfect process owing to the multitude of uncontrollable variables belonging to the rockmass, the geometry of the blast, the impact of previous blasts on free face(s), the explosive imperfections and performance uncertainties, role of delays and their malfunctioning, and last but not the least, the human factors involved in the execution of a blast. The human factors include error in drilling, explosive loading, stemming and wrong firing of the blast, and inability to identify or foresee situations of blast design or rockmass that have a high potential to generate flyrock. But then the premise of perfect engineering does not apply as the conditions are better related to a mistake and not to engineering. The situation thus becomes paradoxical, as either a blast is engineered or not is a binary. Hence, the use of all the blast design variables for flyrock distance prediction needs to be examined.

A well-accounted blast means that blast design has been perfectly matched to the rock and proper protocol for explosive selection has been followed. There are evidences that the blast design variables, including rockmass characteristics, do not match exactly with the calculated values but present a distribution (Thornton et al., 2002). It is therefore essential to understand that there will be variations in blast design variables and hence the results, thereof. If the values of such variables do not fall in acceptable limits in practice, flyrock is bound to occur, along with other side effects like increased ground vibrations and air overpressure. Since due diligence is done to design and implement a blast design, prediction of flyrock, fully based on blast design variables, needs further evaluation.

3.2 BLASTHOLE DIAMETER AND FLYROCK

Drill or blasthole diameter can be considered as a system constant because its diameter is practically fixed, unless otherwise specified, for a particular blasting operation. However, a large range of blasthole diameters, e.g. 0.032–0.310 m, are deployed in field as documented in several reports of flyrock distance prediction. Blasthole diameter is one of the fundamental design considerations in blasting as all blast design variables, for a particular rockmass, are correlated to it, as has been explained in Chapter 1. The fundamental work of Lundborg (1974) employed blasthole diameter in flyrock distance prediction and did not include blast design variables for prediction, despite some problems in their modelling method (Hustrulid, 1999). Despite that, Lundborg (1974) presented a simple deduction of flyrock distance that has found support in standard texts (Dowding, 1985; Persson et al., 1994). The criterion represents the worst-case scenario in both block and actual field conditions.

Drill or blasthole angle, particularly in the case of flyrock, has a strong influence on the direction of the trajectory. There are very few studies on the subject like case

TABLE 3.1
Results of Simulation of Flyrock Distance (R_f) with Change in Blasthole Angle

Case	Blasthole Inclination (°)	Initial Velocity (m/s)	Launch Angle (°)	R_f (m)	Difference (m) from vertical (Case 1)
1	90	100	45	839	-
2	85	100	40	834	5
3	80	100	35	806	33
4	75	100	30	755	84

study (O'Meara, 1994) and mention of role of angled blastholes on flyrock (Bajpayee et al., 2004; Rad et al., 2020), computer application (Favreau & Lilly, 1986) etc. A detailed account of the trajectories involved in cast blasting and flyrock is provided by Chernigovskii (1985) who concluded that with the change in inclination of a blasthole, the throw can be oriented in a particular direction. The direction of the movement will thus change from horizontal, in the case of vertical blasthole, by the angle of inclination of the blasthole. Assuming that a flyrock is projected at 45° from a vertical blasthole and if the orientation is changed by 10°, the flyrock will be launched at 35°.

With the help of a simple apps (Kim, 2022; Ranganath, 2022), we can calculate the difference in flyrock distance. For the aforementioned example, assuming a velocity of 50 m/s with a mass of 0.5 kg and drag coefficient of 1.4 on a level ground, with an initial launch angle of 45° in the case of a vertical bench. The simulation of flyrock distances (Table 3.1) over a change of 15°, in an interval of 5° from the normal, results in a difference of flyrock from 5 to 84 m.

The exercise in Table 3.1, which can be extended to other conditions of mass and drag coefficient values, thus demonstrates the importance of blasthole inclination in flyrock distance predictions.

In addition, the detonation velocity (*VoD*) of commercial explosives, like ammonium nitrate—fuel oil (ANFO) and slurry—increases logarithmically from 1500 m/s to 4500 m/s, for 0.050 m to 0.15 m, and becomes asymptotic beyond 0.150–0.250 m diameter blastholes (Ash, 1973). The VoD of the ANFO can be approximated (Ash, 1973) with Equation 3.1:

$$c_d = 2133\ \ln(d) - 8674; \left(0.05 \times 10^2 \le d \le 1.5 \times 10^2\,\text{m}\right) \tag{3.1}$$

where d is the blasthole or explosive in m and c_d is the detonation velocity in m/s.

Other works on dependence of *VoD* on charge diameter include tests on ammonium nitrate and activated carbon mixtures in steel tube (Miyake et al., 2008); mass fraction of explosive products in ammonium nitrate and fuel oil mixture (ANFO) using an Eulerian hydrocode, and compared with the experimental results from steel tube tests (Kinoshita et al., 2011); emulsion explosive trials (Mertuszka et al., 2018); gelled explosive sensitized with glass micro-balloons (Higgins et al., 2018).

The role of drill diameter can be thus summarized as follows:

1. Blasthole diameter (d) has a direct role in blast design and any mismatch of d with rockmass can result in unwanted blast outcomes like flyrock.
2. Higher diameter blastholes results in higher detonation velocity of the explosive for most common diameters.
3. Higher detonation velocity means higher blasthole pressures.
4. Higher blastholes pressures will impart higher impact on rockmass and flyrock, if generated, and thus, have higher propensity of generating flyrock.
5. Blasthole inclination will impact the post-release behaviour of flyrock as it has a bearing on its launch angle.

However, the role of explosives and its properties vis-à-vis flyrock is not well documented.

3.2.1 Specific Charge

The specific charge (q) in surface blasts varies for different kinds of rocks from 0.2 to 0.9 kg/m^3. The specific charge can be correlated to the RQD according to Pal Roy and Dhar (1993), to the rock quality index (Leighton, 1982), or to the drilling index (Jimeno & Muniz, 1987). The specific charge can also be correlated to seismic wave velocity in the field (Broadbent, 1974; Muftuoglu et al., 1991; Tripathy et al., 1999). However, King et al. (1988) contended that the specific charge, the most common design factor in blasting, does not provide a good method for blast design. King et al. (1988) are right in their claim as specific charge doesn't tell anything about the explosive and its properties, except for the quantity of explosive used, which in turn have a significant impact on the blast outcome (Raina & Choudhury, 2014).

Irrespective of the role of specific charge in fragmentation, higher values of q tend to produce flyrock. This has been demonstrated by Lundborg et al. (1975). Hence a perfect match of specific charge is expected from the designers and blasting personnel to have a control on flyrock.

Specific charge, despite its shortcomings, is one of the design factors that has a direct correlation with the heaving and throw of the blasted muck. Excessive specific charge has been correlated with the excess throw of material and even flyrock distance. However, use of its constituent variables like burden, spacing, drill length, or bench height and explosive per hole for flyrock distance modelling has not been explained so far. Such correlations developed have a shortcoming of using average values of all the blast design variables while as only extreme values will produce flyrock.

3.3 ROLE OF STEMMING IN FLYROCK

Although a design aspect, stemming requires special attention since it has been listed as a major cause of flyrock. Generally, a portion of top of a blasthole is left uncharged and is filled with an inert material like drill cuttings and the length is called stemming length. The first detailed account of stemming and its various aspects can be found in Coulibaly (1982). While drill depth, burden, and spacing are single-dimension variables, stemming has three elements, viz. stemming length, stemming type, and method of stemming. As the term "stemming" has been used, without any qualification in lot of studies, as length of stemming, the other two aspects of stemming mentioned earlier are equally important.

The main purpose of stemming is to provide proper confinement of explosive gas pressures, over a short span of time, to aid in fragmentation (McHugh, 1983), throw, and to control unwanted effects of blasting like air overpressure and vibrations. Konya and Davis (1978) concluded that the performance of explosive is dependent on retention of the explosive pressures within the blastholes. In an experiment with no stemming and stemming (Zhang et al., 2020), it was observed that the p-wave leading the compression and tailing tension pulses were 20% and 50% higher, respectively, in the case of stemmed explosive. Zhang et al. (2020) also reported increase in tailing pulse by 30% and five times increase in crack radius at crack arrest in stemmed explosive. It is demonstrated that the confinement of explosive energy owing to proper stemming for each millisecond of increased retention time increases energy efficiency of blasting and aids in fragmentation (Eloranta, 1994).

3.3.1 Stemming Length

The length of stemming should be sufficient to result in proper breakage and flyrock control. There are no set rules for stemming length (Oates & Spiteri, 2021). The authors (Oates & Spiteri, 2021) further stated that crushed aggregate with a stemming length of 20–30 × d and 0.7 × burden is generally recommended in literature for small and large blasthole diameters, respectively. Other studies report that stemming lengths may vary between 16 × d and 48 × d (m) (Gustafsson, 1981; Langefors & Kihlstrom, 1978) or 20 × d and 60 × d (m) for soft and hard rock, respectively (Jimeno et al., 1995) and a length of 25 × d should be maintained wherever possible. However, Schwengler et al. (2007) recommended a stemming length of 24–34 × d.

Pradhan (1996) concluded that stemming length varies from 12 times the diameter of the blasthole in the case of hard competent rock (uniaxial compressive strength, σ_c >210 MPa) to 30 times the diameter of the blasthole for soft competent rock (σ_c > 30 MPa). A general rule of thumb is that stemming length should be 0.7 × B with a value of 1.0 × B for flyrock control. However, it is known that stemming length is directly proportional to the burden and the constant of the equation may vary from soft to hard rock, with a value of 0.67 × B reported by Ash (1973). Konya and Konya (2018) provided a comprehensive account of the stemming length, material, and loading while providing a justification for the rule of thumb, i.e. l_s = 0.7 × B through calculations. The loss of energy to the tune of 50% in unstemmed holes or lack of proper packing of the stemming material is documented by Brinkman (1990). Also, it has been observed that higher lengths will produce coarse fragmentation (Figure 3.1a), higher vibrations and air overpressure, and lower stemming lengths may result in flyrock (Figure 3.1b).

Rai et al. (2008) conducted tests in sandstone strata and concluded a stemming length of around 0.6 × B yielded optimum fragment size. The effect of stemming

FIGURE 3.1 (a) Boulders produced in a no stemming blast. (b) Flyrock due to least stemming.

length-to-burden ratio was demonstrated by Rathore & Lakshminarayana (2006) with a low degree of correlation in the data. Adhikari (1999) contended that in a blast design, stemming length was one of the most neglected variables. It was determined that the flyrock did not exceed 100 m at a stemming to blasthole diameter ratio of 20 and was the best measure, along with the use of angular granular material, for controlling flyrock in limestone mines.

Trivedi et al. (2015) conducted a sensitivity analysis of blast design variables through a neuro-fuzzy inference system (ANFIS) and found that the stemming length is the most sensitive variable for controlling flyrock in large-diameter blastholes. Shahrin et al. (2019) conduced numerical simulation on the effect of stemming to burden ratio on rock fragmentation by blasting and reported that the mean particle size increases with the increase in stemming to burden ratio, as also supported by Chakraborty et al. (2002).

Since burden is related to blasthole diameter and stemming is a function of burden, with a range depending upon the rock type, there is a case of optimization of stemming that provides best fragmentation. However, wherever flyrock is a concern, optimization may not work as flyrock prevention requires longer stemming that may not favour optimal fragmentation. This is the area of interest that practically leads to compromise in fragmentation, if flyrock is an issue at a mine.

The extreme conditions of flyrock from stemming zone are explained through "scaled depth of burial" (B_{sd}) concept (Chiappetta et al., 1983; McKenzie, 2009) that is derived from cratering experiments. The B_{sd}, as further explained in Chapter 4, is defined as the length of stemming plus half the length of charge contributing to the cratering effect, divided by the cube root of the weight of explosive contained within the portion of charge contributing to the crater effect. This concept has a bearing on length of stemming and hence flyrock.

It is hence evident that the length of stemming has an influence on flyrock generation. This is the critical area of the blasthole where there is a tendency of cratering and if that happens, flyrock is imminent. If the length of stemming is equal or more than burden, then there are least chances of flyrock.

Several conditions of blasts arise in which flyrock from stemming zone may emerge out of the blast face. In an ideal case of blast where stemming zone is not

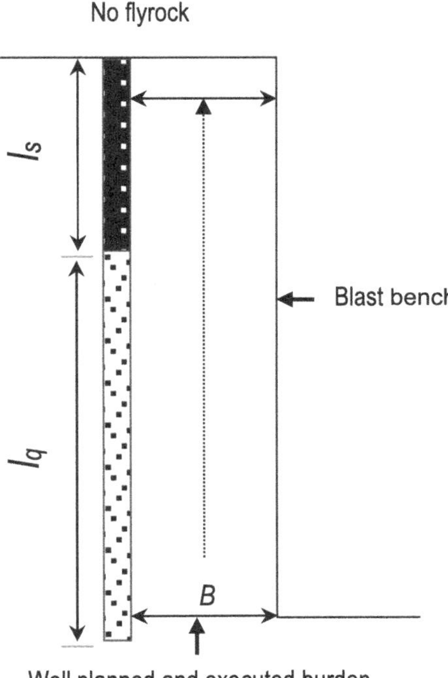

FIGURE 3.2 Ideal bench with proper burden and stemming length. (B is the burden, l_s is the stemming length, and l_q is the charge or explosive length)

disturbed due to earlier blasts, the stemming length is proper and as per design, there should be no chances of flyrock (Figure 3.2).

A peculiar situation in the case of ideal situation can be that of the damaged bench top owing to previous blasts. In this case, the stemming zone of a planned blast lies in a damaged rockmass and presents a case for flyrock generation, even if all other blast design variables are as per plan (Figure 3.3).

However, if the stemming length is less than the planned in such a case, there is every chance of flyrock generation (Figure 3.4) in every possible direction from the top of the bench. In this case, the top of the blast face becomes the path of least resistance for the explosive gases, hence providing perfect conditions for ejection of flyrock from the stemming zone. This condition presents a typical case of cratering with least depth of burial of the explosive.

Another instance of flyrock from stemming arises out of uneven face conditions due to previous blasts. In such cases, the crest of the first-row holes of the blast has low burden than designed. Flyrock can occur in this case if designed stemming length is applied, thereby resulting in high explosive concentration near the stemming zone. Thus, flyrock can emerge towards the top of the bench (Figure 3.5) with more tendency of flyrock occurrence towards the crest of the free face.

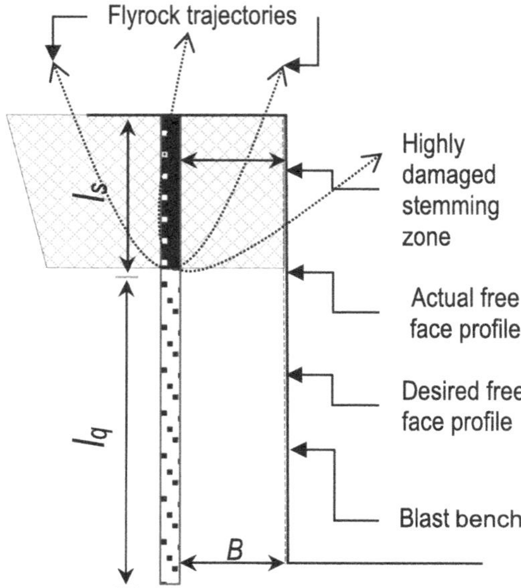

FIGURE 3.3 Ideal burden and stemming but the stemming zone is damaged that can result in flyrock. (*B* is the burden, l_s is the stemming length, and l_q is the charge or explosive length)

A combination of the damaged top and front of the blast face can occur resulting in a very peculiar situation. In such a case, the flyrock will occur in all directions with greater propensity towards the free face, as shown in Figure 3.6. The stemming zone may not be able to hold the stemming for longer duration for burden to break properly and the high-pressure gases to release through fragmented face. Movement of the blasted rockmass towards the front may even be impeded resulting in tight muck also.

The aforesaid conditions of stemming can occur even in multiple row blasts that will have greater tendency to crater owing to less relief of the blastholes in the back rows. However, such conditions cannot be a design consideration, but are errors of engineering judgement or implementation. If blasts are conducted in such conditions without due diligence, it will not be improper to call these as mistakes.

3.3.2 Stemming Material

Stemming types practiced in mining and other areas are varied in nature. Significant experimental investigations on stemming material and its influence on blast performance have been documented through times immemorial. One can trace such citations even to 1912 (Snelling & Hall, 1928)[1] who conducted experiments on lab scale to determine the effectiveness of stemming in blast performance—it was determined that clay worked better than other substances used as stemming. A detailed review

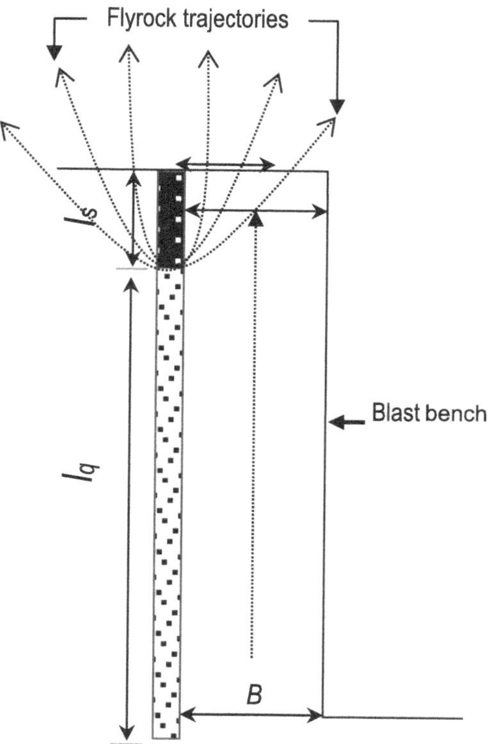

FIGURE 3.4 Well-planned blast, least stemming length. (B is the burden, l_s is the stemming length, and l_q is the charge or explsive length)

of classical works is provided by Coulibaly (1982) who conducted tests on different stemming materials like drill cuttings, tailings, mixture of sand and clay, and water-filled plastic bags, to ascertain their impact fragmentation, vibration, and noise levels. Coulibaly (1982) reported that their tests were more or less in confirmation with the findings of Joslin (1982).

Armstrong (1994) devised a method to test the performance of stemming and conducted laboratory and field tests and concluded that 1/10th × d size of stemming particles showed best results. It was believed that such grain sizes have an interlocking mechanism and prevent gas energy from being released through the stemming zone. Tang et al. (1998) determined that sand and chips, used as stemming material, performed better than other materials. Dobrilović et al. (2005) recommended 16/32 inch fraction (approximates to 0.15× d) of stemming material for all cases.

Rustan et al. (2011) mentioned:

Principally the size of the stemming material should be as large as possible but less than 1/3 of the diameter of the hole, to avoid pieces getting stuck

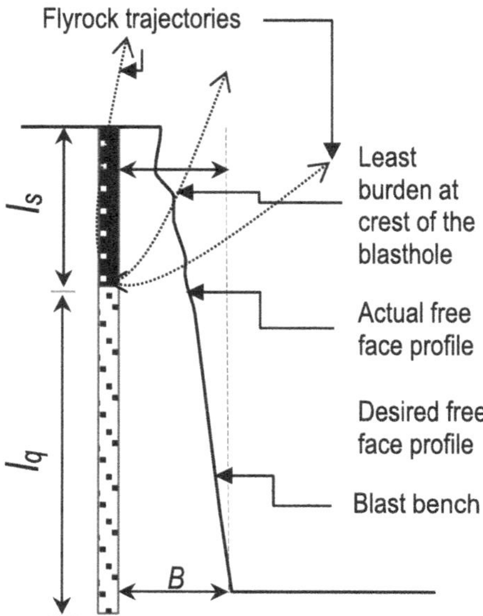

FIGURE 3.5 Uneven and less burden at the crest of the blasthole. (B is the burden, l_s is the stemming length, and l_q is the charge or explosive length)

in the hole when filling the blasthole, and to obtain the maximum bridging effects at blasting. Too large pieces of stemming material may damage the initiation material in the blasthole and should therefore, be avoided. In coal mines it is preferable to have fine material because of the risk of fire. For large hole diameters the shape of the rock pieces used for stemming should be angular. The length of the stemming should be about the same size as the burden.

Kim and Lee (2011) compared performance of crushed granite sand, sea sand, river sand, clayed soil, and water as stemming material in model experiments, and determined that the axial strain was maximum with crushed granite sand and stemming ejection velocity was minimum with water stemming. Kim and Lee (2011) further found that the axial strain and stemming ejection velocity show inverse relationship. Cevizci (2019) compared the performance of plaster and conventional drill cuttings as stemming material, using numerical modelling, and that the maximum pressure at top of the explosive column was almost 1.625 times higher in the case of plaster. Vasconcelos et al. (2016) used conventional and non-Newtonian fluids as stemming material in several tests and reported that the air overpressure was reduced by 84% and 26% in the case of no-stemming and sand-stemmed coupled blastholes, respectively. Non-Newtonian mixture used as stemming (Vasconcelos et al., 2016) was investigated and results indicated that the material provided better confinement resulting in reduced air overpressure and

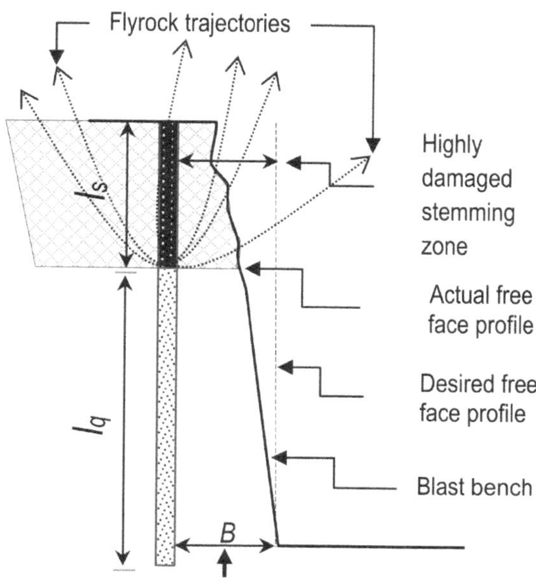

FIGURE 3.6 Broken top blast face and low burden at the crest of the blasthole creating typical situation for flyrock occurrence. (*B* is the burden, L_s is the stemming length, and L_q is the charge or explosive length)

ground vibration, and increase of the strain propagation in the rock. Such materials are better than sand stemming as observed by Ko et al. (2022) in full-scale blast experiments.

The role of stemming material in flyrock occurrence: Oates and Spiteri (2021) claimed that there was no or scant literature supporting role of stemming material on flyrock occurrence. Schwengler et al. (2007) used high-quality angular stemming material, a by-product of concentrating process of 0.06–0.014 m size and least fines ($10\% \times d$) in 0.115–0.152 m diameter blastholes, and reported reduction of blast danger zone to 250 m from 600 m. Sharma and Rai (2015) reported that throw of fragmented material was 21.2% higher in blasts stemmed with drill cutting as stemming material and vertical cratering was greater in aggregate of crushed stemming material. Sharma and Rai (2015) found that in the case of blastholes stemmed with crushed aggregate, flyrock did not extend beyond the bench floor, whereas holes with stemming of drill cuttings generated flyrock that reached other benches. Such findings are also supported by other works, e.g. Adhikari (1999) and Sazid et al. (2011).

In nutshell, the role of stemming material is not confirmed to have a direct relationship with flyrock. There are conditions in which stemming may eject prematurely due to its improper nature and result in flyrock. However, unless the conditions mentioned in Figures 3.5 and 3.6 are met, there are little chances of flyrock. Even if flyrock is generated, it will be aligned in a direction parallel to blasthole and can reach higher lengths but cannot be thrown to a great distance owing to its near 90° launch angle in vertical blastholes. Such flyrock can be dangerous only to high-placed objects like overhead powerlines, which are not generally encountered in mines. However, in the case of inclined blastholes, the conditions may be different, wherein flyrock from collar of the blasthole or stemming zone can travel significant distance and damage nearby objects of concern.

3.3.3 Stemming Method and Practice

The stemming methods are varied in nature that range from manual mode to mechanical devices. Manual method consists of loading the stemming zone with inert materials as mentioned earlier. There are two types of practices followed in such case, viz. simply filling up the stemming length with inert material by shovels and filling with tamping of the stemming material with a stemming pole or tamping bar of wood or similar material. The latter method is better as it ensures compactness of the stemming and provides sufficient confinement to the explosive gases and prevents their premature release. This ensures movement of the broken rockmass towards the front instead of top of the bench.

Konya and Konya (2018) say that proper stemming not only increases explosive efficiency but also helps to reduce flyrock. However, they stressed that stemming should be loaded properly in the blasthole if there are chances of mixing of stemming materials with the explosive. The influence of stemming method on ground vibration and air overpressure is documented (Mpofu et al., 2021) and there are casual references to the methods, but no specific mention of control of flyrock is given in such publications. Mpofu et al. (2021) in their studies determined discrepancies in manual recording of data and need for automatic systems for such data acquisition and recommended use of best practices for stemming.

Mechanical devices[2] are also used to stem the blasthole after charging it with explosives. McNail et al. (1999), through an extended work of Wilkins (1997), reported such devices for underground inclined hole stemming. Such device is mounted on a truck and collects stemming material and fills the blasthole. However, there is little or practically insignificant information available on such devices used for stemming in surface mines.

3.3.4 Stemming Plugs

One of the recent advances in stemming is the application of devices called "stemming plugs." These are essentially rubberized or plastic conical gadgets that come in various shapes and sizes. A detailed account of such devices can be found in Maram and Uttarwar (2020). Three such plugs were tested by Lubbe and Frye (1998) for different diameter blastholes and observed that stemming ejection velocity drops faster in the case of stemming plugs when compared with the rock chips stemming type. Also, in a particular type of stemming plug, Lubbe and Frye (1998) reported increase in heave, and a bypass of delay when such plugs were used as decking in comparison to rock chips. Other studies on use of stemming plugs and their results are summarized in Table 3.2.

TABLE 3.2
Results of Some Studies on Use of Stemming Plugs in Blasting

Year	Citation	Plug Type	Results
1993	Worsey (1993)	Conical cone with stabilizing rod	Substantial reductions in stemming ejection velocity/ejection elimination, stemming length reduced by up to 35% over conventional. p_{oa} reduced by 8–25 dB. Inverse relationship between rock face movement and stemming ejection velocity derived
2004	Correa and Navarrete (2004)	Polyethylene, polyvinyl chloride	Stemming retention times varied from 269 ms with Brand-A plug to 364 ms with Brand-B plug with a 95 ms difference in their average values at same stemming length, while between Brand-B and drill cuttings there is a 47 ms difference (not within the 95% confidence interval for the range of retention times with Brand-B). A reduction in fragment size by 22% in 75% passing was recorded
2013	Choudhary and Rai (2013)	Rubberized polyvinyl chloride	Use of stemming plug in the blasthole below the stemming column produced favourable results
2016	Neves et al. (2016)	Not known	Air-deck with stemming plugs yields better fragmentation in comparison with drill cuttings
2019	Rehman et al. (2019)	Plastic, air, and cement mortar	Performance of three types of plugs evaluated through digital image analysis of fragmented rock. The plastic moulded stemming plug was found to be the best among all selected plugs
2019	Bhaskar et al. (2019)	Plastic funnels plug from local market	Funnel-shaped stemming plugs controlled flyrock and increased the velocity of detonation due to wedge effect, thereby improving fragmentation as well as mine economics. The information is however qualitative only
2021	Rehman et al. (2021)	Plastic, air, and cement mortar	Used three types of stemming plugs, viz. Plug 1 (plastic moulded), Plug 2 (air-plug), Plug 3 (cement mortar), and reported improvement in overall mine economics by improving fragmentation and translating into savings of US$ 0.03–0.045/ton of limestone and controlled flyrock. However, no data on flyrock is provided
2022	Choudhary and Agrawal (2022)	Rubber	Reported reduction of fragmentation by 53%, backbreak by 56%, and flyrock distance by 61%

Although there are some safety issues with the use of stemming plugs (Hartmann et al., 1950), there are studies that favour the hypothesis of improvements in blasting results in terms of fragmentation and heave.

In view of the discussions on stemming length and material used for stemming, it is imperative that the retention of blasthole pressure is of prime importance to achieve greater efficiency of explosive energy. A possibility exists about retention of blasthole pressure with stemming plugs for a few microseconds or a millisecond, that may be sufficient to drive further fragmentation and favour movement of the fragmented rock towards the front of the blast and hence preventing venting of gases and flyrock from the stemming zone. However, the control of flyrock with stemming plugs needs to be properly documented through comprehensive studies that are currently lacking.

3.4 ROLE OF BURDEN IN FLYROCK

Burden is one of the most important blast design variables that influences flyrock distance. Burden, however, has a direct relationship with drill diameter. The optimum breakage burden (B_{ob}) can be estimated by the Livingston (1956) crater theory (Equation 3.2) in terms of total charge mass (Q) that were further defined in terms of burden by Dick et al. (1990) and Da Gama (1984) as in Equations 3.3 and 3.4, respectively.

$$B_{ob} = k\sqrt[3]{Q} \tag{3.2}$$

$$log\sqrt{\left(B_{ob}^{2} + r_{a}^{2}\right)} = 1.846 + 0.312logQ \tag{3.3}$$

$$r_c = \sqrt{\left[3B_m^2 - B^2\right]} \tag{3.4}$$

where r_c is the crater radius (m) in rock, B_{ob} is the optimum breakage burden distance or charge depth in m, k is a constant of proportionality expressing rock and explosive properties, Q is the mass of explosive in kg, and r_a is the apparent crater radius and the equivalent TNT charge mass (Q).

Accordingly, an optimum burden will thus produce the desired results in terms of fragmentation and throw. However, two conditions of the burden, i.e. lesser than optimum and more than optimum, arise. In the case of low burden, the throw will transform to excessive throw and even flyrock travelling towards the free face. However, in the case of excess burden than optimum, the confinement of the explosive will be greater, and thus may induce flyrock from the top of the face that can be launched in any direction, including back of the free face. These may be the situations in several studies listed in Chapter 4, where flyrock distance has been modelled with burden as a variable.

3.5 IN SITU ROCKMASS CHARACTERISTICS

In situ rockmass nature is an important uncontrollable factor in blasting. It not only effects the fragmentation and heave, but also has a strong influence on the vibrations, air overpressure, and flyrock. There are several conditions of rockmass that may influence the flyrock occurrence. The confinement capacity of the rockmass is determined by its condition or impedance characteristics and hence flyrock estimation should consider geology and its structure over a complete spectrum (Blanchier, 2013). Mohamad et al. (2013b, 2016, 2018) provided a comprehensive detail of influence of rockmass characteristics on flyrock and concluded that less mean joint spacing with larger apertures resulted in excess flyrock (whether number of flyrocks or greater distances are not specified).

However, it may be noted that larger apertures in competent rockmass may result in flyrock, but logically, lesser joint spacing or highly fractured rock tends to release the gas pressures quickly through the joints and hence there is less chance of flyrock to reach far distance, provided the blast design was engineered properly for that type of rock. Rockmass with high joint frequency is bound to have smaller fragments, which in turn have less probability of being projected to large distances. In the case of massive rock, the degree of confinement of explosive gases is high, which results in significant built-up of gas pressure, which means higher time of application of force on fragments. Under such conditions, there is a probability of larger fragments travelling greater distances.

Flyrock occurs in all types of rockmass, but it is necessary to understand that flyrock can occur in extreme conditions only. Such an issue was raised by the author (Raina et al., 2015) to help the community to develop rational models for flyrock. This involved a comprehensive scheme that included rockmass conditions responsible for flyrock generation as explained further.

3.5.1 Rockmass Conditions and Flyrock

Several conditions of blast arise when, even after proper engineering, flyrock is generated. Most of the times, these are related to the weakness planes, as supported by significant studies (see Chapter 6), in the rockmass which are not known or may be overlooked during blasting. A description of conditions that lead to flyrock generation include the presence of the following:

1. Weak strata in competent rock formation of the blasting bench. This may include presence of the strata at different places across the competent formation or at different levels of the bench. Such weak strata can be horizontal (Figure 3.7), dipping towards or out of the face (Figure 3.8) or dipping into the face (Figure 3.9).

All the aforementioned conditions will produce flyrock. However, out of the three conditions provided in Figures 3.7–3.9, the presence of weak strata dipping into the face can be very dangerous as the launch angle in such cases can match the condition that favour maximum projection of the flyrock. Such rockmass conditions combined with other blast face disturbances like weak stemming zone, damaged face, and damaged stemming zone are potential reasons for the flyrock occurrence.

A variation of such a condition, that has been seen to be very critical for flyrock generation, is occurrence of clay bands of secondary origin in vertical direction in wide and open joints in a limestone mine (Figure 3.10). In this example, an error of judgement led to severe flyrock event wherein several flyrocks were projected through the clay band and hit OCs situated far from the place of blasting.

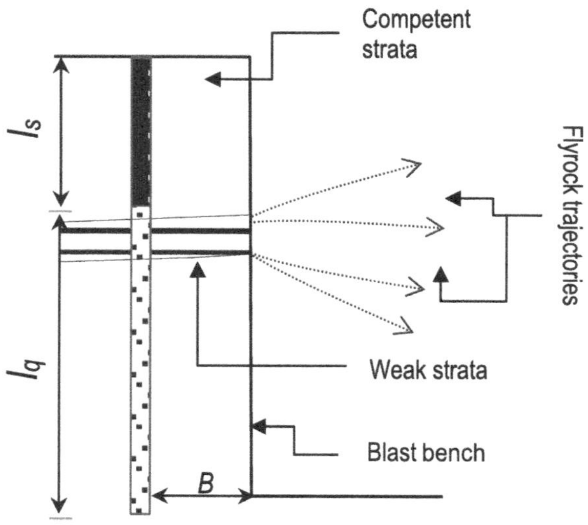

FIGURE 3.7 Weak strata in competent formation in horizontal position. The weak strata can occur at any position in the blast bench and the nature of flyrock will vary accordingly. (*B* is the burden, l_s is the stemming length, and l_q is the charge or explosive length)

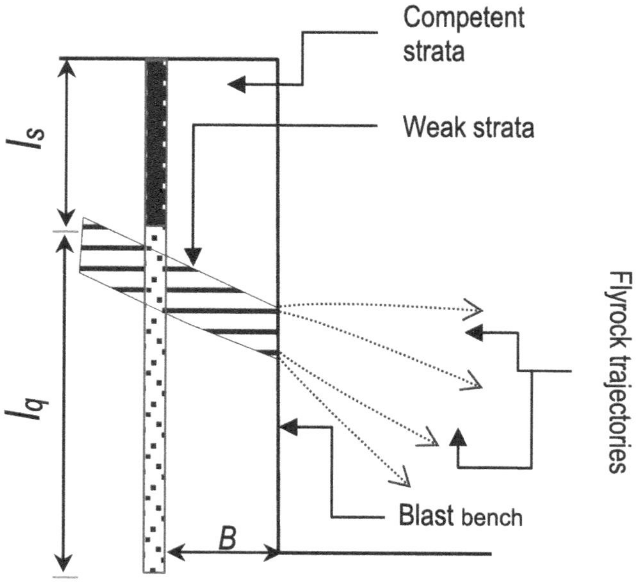

FIGURE 3.8 Weak strata in competent formation dipping out of face, and can occur at any place in the blast bench. (*B* is the burden, l_s is the stemming length, and l_q is the charge or explosive length)

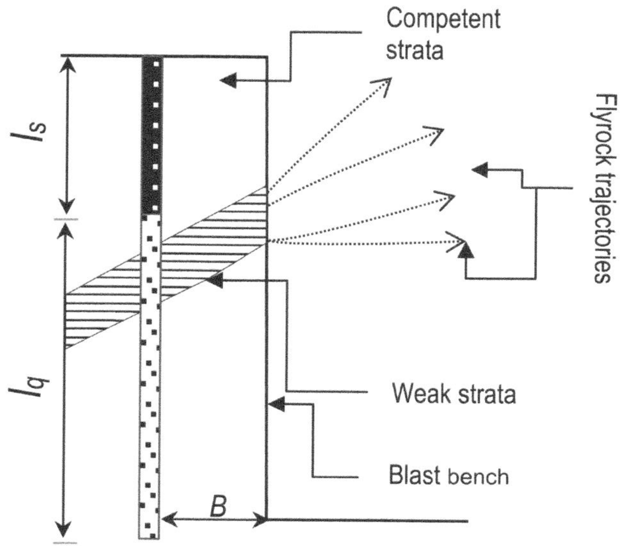

FIGURE 3.9 Weak strata in a competent formation dipping into the face and can occur at any place in the blast bench. (B is the burden, l_s is the stemming length, and l_q is the charge or explosive length)

The aforementioned case details are listed as follows:

Blasthole diameter:	110 mm
Rock type:	limestone with clay bands dipping in vertical into the blast face
Explosive type:	cartridge emulsion of 83 mm diameter
Blast design:	bench height 8 m, burden 3.5 m, spacing 4.5 m, total charge per hole of 45.3 kg, stemming 3.2 m, stemming-to-burden ratio 0.9
Firing pattern:	diagonal hole to hole
Firing timing:	17/25 ms per delay with in-hole delay of 400 ms
Flyrock direction:	80° with respect to face
Throw:	more than 50 m
Flyrock:	21 numbers thrown at distance ranging between 75 and 250 m
Flyrock size:	0.23 m projected to a distance of 250 m and hit an industrial shed

During further permitted trials, the blast design was modified in this case that included prior assessment of rockmass and weak zones, increase in stemming length, use of crusher chips as stemming material, and deployment of conveyor belts over zones of weakness the competent rock. Around 110 blasts with the modified design were monitored with the help of video camera for flyrock monitoring. None of these blasts generated flyrock despite several issues with the rockmass.

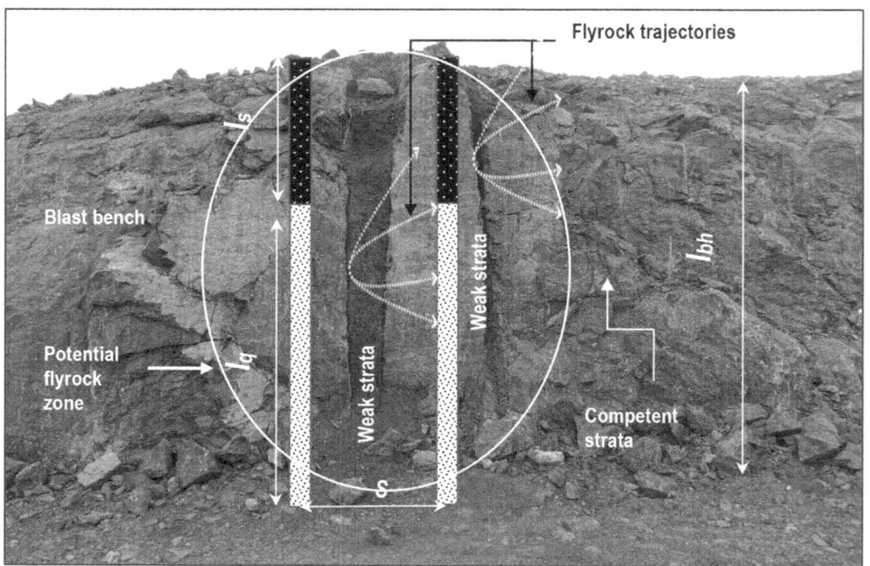

FIGURE 3.10 Presence of clay band in competent limestone rockmass—a situation highly conducive for flyrock generation. (*B* is the burden, l_s is the stemming length, l_q is the charge or explosive length, and L_{bh} is the bench height)

2. Weak planes, structural discontinuities, or open joints in the rock formation of the blasting bench. These are potential areas from which flyrock can emanate. Generally, there is no control over these and many a times, these may not be well-manifested on the surface of the bench.
3. Highly weathered strata in the stemming zone. Such conditions as recorded in several reports are prone to flyrock as there is little or practically no confinement of the high-pressure explosive gases to perform in designed manner.
4. Voids in limestone as is typical of karst topography. This condition is frequently observed in such formations arising out of solution and formation of cavities in the rockmass. If blastholes are drilled through such cavities without logging, these can accommodate extra amount of explosive (Figure 3.11). This condition of a blast is loosely termed as "building a bomb." Such excessive charging can be highly dangerous and render too much of throw of the blasted material alongside flyrock, which can travel several hundred metres.
5. Structural defects and complex geology. In hilly regions, the structure of rockmass is quite complex. This is generally manifest in terms of severe folding and faulting (Figure 3.12). Since the condition occurs throughout the mine, the blaster has no control over the same and are compelled to compromise on fragmentation and throw to control flyrock. There is a possibility of mixed reaction to explosive loading and firing owing to extreme variations in dip of beds in a single blast event.

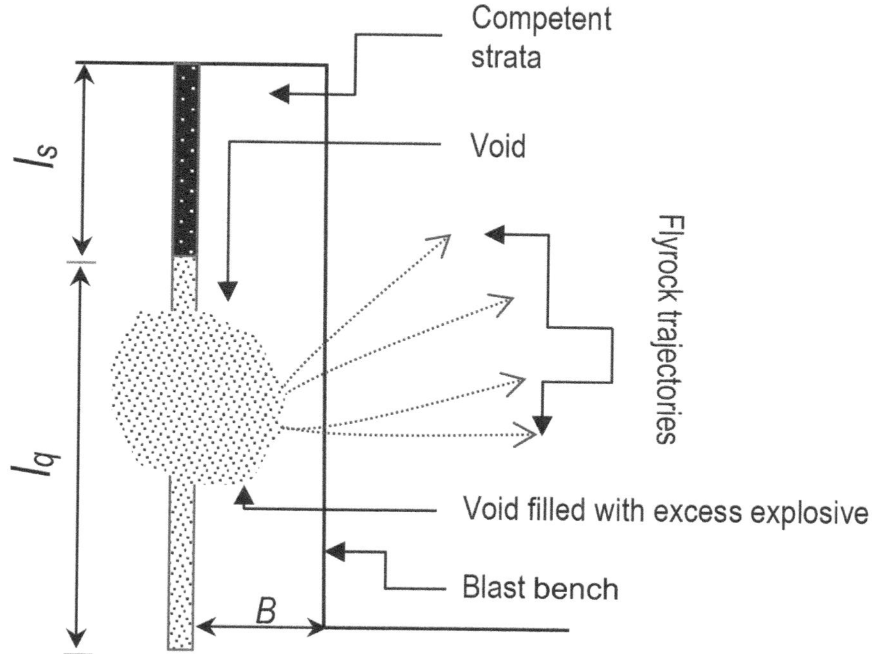

FIGURE 3.11 Presence of void in limestone which can be charged excessively. (*B* is the burden, l_s is the stemming length, and l_q is the charge or explosive length)

The list of rockmass conditions cited earlier favours the observations, over a broad spectrum of reports, that such rockmass defects are conducive for flyrock generation and constitute the prime factors of flyrock causes.

3.5.2 The Unknowns

The unknowns with respect to flyrock include special blasting methods like boulder (Section 2.4) and toe blasting. In both the cases, the criterion of maximum flyrock projection can be applied as there is every chance of flyrock generation. The criterion for maximum flyrock distance should be applied for such cases.

3.6 EXPLOSIVE AND THEIR ROLE

The role of explosives finds a cursory mention in flyrock generation. Major aspects related to explosive concentration (specific charge) and its properties like *VoD* and density have been addressed briefly in Section 3.2. Several aspects of the ideal and mining explosives, although dealt in detail and summarized (Mohanty & Singh, 2012), finally trickle down to *VoD* as commercial explosive are mixtures and behave according to the components and their performance is significantly influenced by the

FIGURE 3.12 Highly complex formation—extreme folding and structural features, presence of mixed competent and incompetent strata.

confinement conditions. Thus, the explosive concentration in a blasthole appears to be a good descriptor of the outcome, though not the best. However, the role of delays requires a better treatment as these define the interaction between the explosive energies unleased at different times in blastholes.

3.6.1 Delay and Its Malfunctioning

Delays are used to provide time gaps in firing of blastholes so that the total charge used in a blast is divided into smaller parts. The charge or quantity of explosive firing in a particular delay is called the maximum charge per delay (Q_{max}). The delayed firing of charges reduces the vibration and air overpressure from a blast that is a function of distance of the blast from monitoring location and Q_{max}. Proper planning of Q_{max} ensures that vibrations are within limits that may cause structural damage. One of the major functions of delay blasting is to provide relief for subsequent holes in a row or between the rows so that proper movement of the broken muck can be achieved and is again dependent on the dynamic properties or rock and explosive (Hustrulid, 1999).

One of the major issues in blasting is delay malfunctioning. This is case with all types of delay detonators, particularly in the case of non-electric detonators with shock tube combination (NeSt). Such delay detonation systems like NeSt have an in-built delay of few hundreds of milliseconds (also called in-hole or down-the-hole delay, DTH) to enable initiation of all detonators in a blast that finally detonate at times designated by surface delays. This helps to avoid cut-off due to fragments from previously fired holes falling on trunklines and is mostly observed in firing with detonating cord. However, the delay elements have a scatter in firing timings (Agrawal & Mishra, 2021; Silva et al., 2018). Accordingly, three types of delay malfunctions are identified that may be of interest insofar as flyrock is concerned.

1. Delays are connected in a row-to-row pattern and a delay element in front does not fire, while other blastholes fire in subsequent rows (Figure 3.13) owing to delay scatter or malfunctioning. This situation can be quite dangerous as the burden in the case of second or subsequent rows adds up, produce circumstances for perfect cratering, and conditions that result in flyrock.
2. Delay element (DTH) scatter is too high and blastholes do not fire at designated times. In this case, some holes may fire before or after

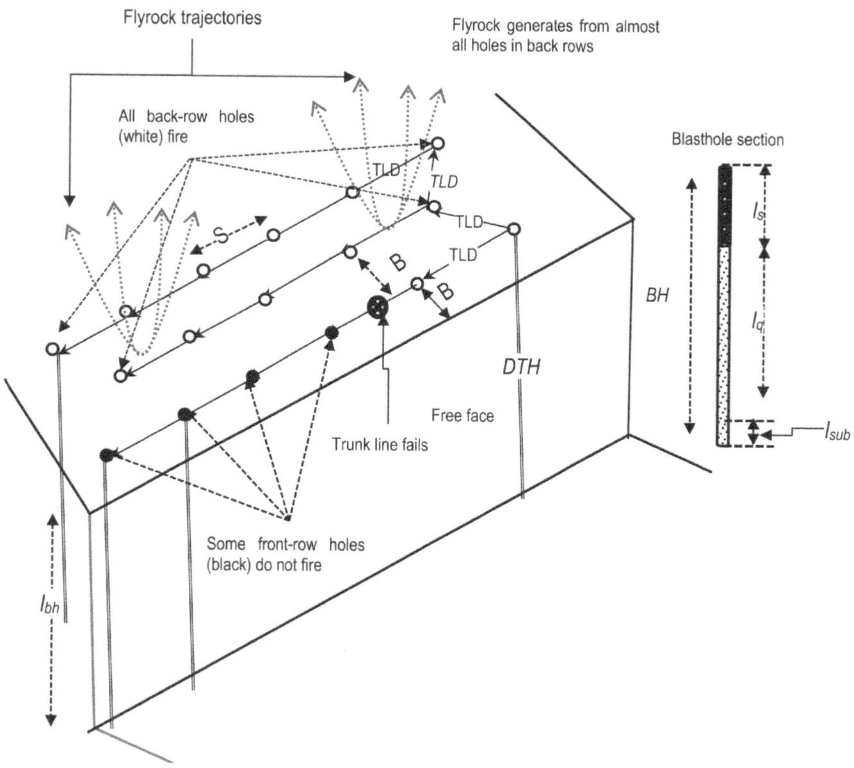

FIGURE 3.13 Delay malfunction in the first row of blast—the holes thereafter in the row do not fire, while other rows fire leading to high confinement of back row holes and hence flyrock generation. (*B* is the burden, *S* is spacing, l_{bh} is the bench height, *BH* is blasthole, l_q is the charge length, l_s is the stemming, *DTH* is down the hole detonator or delay, *TLD* is the trunk line delay), l_{sub} is sub-drill length

FIGURE 3.14 An example of delay malfunction due to scatter resulting in flyrock. (a) Blasthole in later sequence fired before others in queue. (b) Flyrock(s) emerging from the blasthole by cratering due to over-confinement.

the designated time and delays firing randomly can result in over-confinement of blastholes and generate flyrock. A case of a coal mine blast (Figure 3.14) in which a delay in the front row blasthole fired earlier to other holes. This resulted in the over-confinement of the blasthole and resulted in several flyrocks.

3. Another case relates to less delay for back-row blastholes in a blast that restricts movement and relief of blastholes in front rows. In absence of a free face or relief, the blast has a higher tendency to crater and produce flyrock.

3.7 FLYROCK DOMAINS

As introduced in Chapter 2, flyrock presents a very complex topic where its different aspects fall in different domains, as explained in Figure 3.15. Three major domains are identified in the analysis of flyrock, viz. mine, air, and landing surface.

The mine domain incorporates rockmass with its defects, explosive including all imperfections in the blasthole connectors, and finally the rock–explosive interaction that provides impetus to the flyrock. This defines the time of exit, the exit or initial velocity, the launch angle along with the shape, and weight of the flyrock. The role of blast is limited to this domain only with no further contribution to the travel of flyrock.

The domain of air has several components. Once a flyrock exits the blast face, the forces of wind velocity, wind direction, drag, and translation in air play a major role in defining its course and final travel distance. Unlike ballistic projectiles, flyrock starts retarding under their influence immediately after release from the blast face. This domain practically represents trajectory physics, a well-established science.

Finally, the flyrock on landing can rebound further or can be arrested by local conditions of the surface on which it falls and falls in the domain of landing or rebound. This has already been treated and is defined by several parameters.

Since other domains have been addressed earlier, it will be prudent to deliberate on the post-ejection phenomenon with the help of trajectory physics.

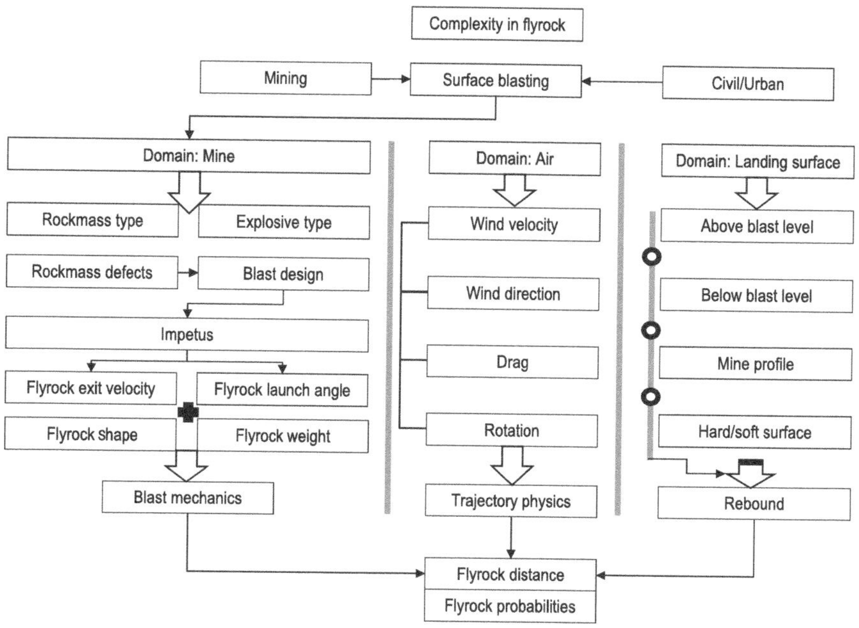

FIGURE 3.15 The complexity of flyrock and its prediction. The figure shows three major domains in which the flyrock can be treated and hence the need for crisp definitions of the independent terms used while treating the flyrock.

3.7.1 Post-Ejection Phenomenon of Flyrock

Flyrock, once released from the blast face, enters an altogether different regime and is governed by principles of trajectory physics. This will be termed as post-ejection phenomenon. The movement of flyrock in air has several aspects that have to be part of the predictive mechanism, when attempted. However, not all such facets of flyrock are considered in studies that can lead to error in prediction (Raina et al., 2015). In order to have an insight into the topic, such aspects are briefly described.

3.7.1.1 Initial Velocity

The velocity of flyrock is one of the basic components that enables prediction of its travel distance. Generally, the velocity is a function of the impetus due to high-pressure explosive gases and equations governing the flyrock distance in terms of velocity will be detailed in Chapter 4. The initial velocity (v_0), that is imparted to the flyrock, is of significance in predictions. The size and density of the flyrock that have been ignored in most of the publications on flyrock is an important constituent that determines the travel of the flyrock in air.

3.7.1.2 Flyrock Angle

The launch angle of a flyrock is quite important along with its initial velocity as these determine the distance of travel of flyrock to a major extent. Low angle of launch, i.e.

10°–30°, will not support the flight of the trajectory to a greater distance and at the same time, there are ample chances that the flyrock will be obstructed and its flight terminated by ground at a short distance.

A flyrock travels far if launched at mid-range angles that may range between 30° and 60°. High launch angles 60°–90° tend to project the flyrock in a vertical direction which do not have much significance as these do not travel greater distances and hence cannot be dangerous, unless otherwise specified.

3.7.1.3 Air Drag

The fragment of rock once released from a blast face is subjected to projectile motion in air. The effect of air drag is dominant as it is directly related to the velocity of a fragment in motion. McKenzie (2009) showed that the kinematic equations overpredict the flyrock distance, as air drag is ignored in calculations.

All bodies moving in gaseous or fluid medium experience a resistance to their movement. The resistive force depends on the viscosity of the medium as well as on the pressure, which develops on the surface of the body due to the deflection and retardation of the flow medium. Viscosity resistance is called as friction drag. The pressure developed due to the retarded flow resistance is called as pressure drag. The effect of each of these components depend on the correlation between velocity, linear dimensions of the moving body, and viscosity. The correlation is characterized by the Reynold's criterion and is given by Equation 3.5 after Chernigovskii (1985).

$$R_n = \rho_{fluid} \frac{k_x}{v} \tag{3.5}$$

where R_n is the Reynolds number of a dimensionless quantity, v is the kinematic viscosity of the medium in cm²/s, ρ_{fluid} is the fluid mass density in $\times 10^{-3}$ kg/m³, and k_x is the size of fragment in m.

Since the viscosity of the medium begins to exert a predominant influence when Reynolds number $R_n < 1000$. The kinematic viscosity of the air is $v = 1.4 \times 10^{-5}$ m²/s and mass density of fluid (ρ_{fluid}) is density of air (ρ_{air}) with average value of 1.29×10^{-3} kg/m³. Results of numerous experiments conducted by Chernigovskii (1985) in a vacuum and other media showed that the viscosity of the medium plays a dominant role for Reynolds number less than 1000 ($R_n < 1000$).

By substituting these values in Equation 3.5, we get

$$\rho_{fluid} \times R_f < 0.014\, m^2 / s$$

Air drag is directly proportional to velocity, and viscosity becomes the determining factor.

$$\rho_{fluid} \times R_f > 0.014\, m^2 / s$$

The effect of viscosity is insignificant in comparison with dynamic head, and air drag is directly proportional to the square of the velocity. Equations 3.6 and 3.7

can be used to estimate the air drag (μ_{air}) or the retardation due to movement of flyrock in air.

$$\mu_{air} = b_d v_x^2 \tag{3.6}$$

where

$$b_d = \frac{C_x \rho_{air} L_x}{2M} \tag{3.7}$$

The drag factor b_d depends on the shape L_x and mass of fragments M in kg and the density of air ρ_{air} in kg/m^3. v_x is the velocity of fragment at a given time in m/s. The drag coefficient C_x (generally designated as C_D) for fragments of different shapes varies between 1.2 and 1.8.

Chernigovskii (1985) further described the criterion for application of air drag in kinematic equations based on Equation 3.2 as follows:

If $\mu_{air} < 0.3$ . . . no correction is needed, if $\mu_{air} > 0.3$ . . . correction for drag is needed

Stojadinović et al. (2013) provided a method to work out the drag coefficient considering ballistic trajectory equations for flyrock for which the terminal velocity of the flyrock fragment is required. Terminal velocity is the velocity of a fragment immediately before it hits the ground. However, the terminal velocity of flyrock is difficult to ascertain since exact position of the landing of a flyrock is not known. One of the studies restricted to steel fragments (McCleskey, 1988) is of particular interest as it reported shapes that could be well representative of rock fragments. The drag coefficients for such fragments reported ranges between 0.497 and 1.484 for different combinations of their weight and velocity. A comprehensive detail of C_D of different shapes occurring in nature for reference is also available online.[3]

The effect of air drag has received its due attention from researchers and continues to be investigated for obvious reasons of non-linear retardation of flyrock fragment in air is well-established for objects of known shapes. A flyrock can, though, assume any shape and result in considerable variation in air drag. The influence of other factors, viz. the translational motion in terms of Magnus effect, as explained further, the wind velocity, and direction and rebound, on the travel of flyrock fragment in air cannot be ignored. Therefore, mere application of air drag may still result in variations in flyrock distance predictions.

3.7.1.4 Magnus Effect

When an object moves in a fluid, it experiences deflection due to the interaction with the medium or when an object spins in a fluid, which is called Magnus effect (Magnus, 1853). Since projectiles from blasting show similar characteristics, Magnus effect should be a case with flyrock also. The Magnus force can be determined by finding the difference of pressure in the front and at the back of the moving object over its cross-sectional area. A simple derivation is provided in Equation 3.8:[4]

$$F_A = \Delta_p A = c_A \frac{\rho}{2}\left(u_1^{\,2} - u_2^{\,2}\right) A \tag{3.8}$$

where C_A is a scalar dependent on the shape and material of the rotating object, u is the speed of the fluid relative to each surface, and ρ is the fluid density.

Although simple solutions of Magnus forces for balls of spherical or known shapes can be obtained from Equation 3.8 (Briggs, 1959), there is no such solution for flyrock as it is an uneven object. The reasons for not having mention of the effect can be that either it is difficult to resolve due to limitations of videography, or there have been no attempts to define or incorporate the same in flyrock distance calculations.

3.7.1.5 Rebound

This has already been described in Section 2.3.

3.7.1.6 Size of Flyrock

Size in terms of its dimensions influences the flyrock distance as the impact will be proportional to the area of the fragment and hence the magnitude of the launch velocity. Once the flyrock is ejected from the blast face, its travel in air is governed by its weight, angle of launch, initial velocity, air drag, Magnus effect, and wind velocity and direction. Hence, size is an important consideration in prediction of flyrock distance. However, in most of the cases reported, the size has not been mentioned. There are only a few citations that relate to flyrock size, e.g. Lundborg (1974), McKenzie (2009), and Mohamad et al. (2013a). The size, the weight, the dimensions, and the shape of the flyrock must be part of any flyrock investigations. The topic is further elaborated in Section 4.8.4.

NOTES

1. Revised version of the original document dating 1912 by Howell and Tiffany (1928).
2. https://patentimages.storage.googleapis.com/60/98/58/daa827bdd515d2/US3961492.pdf
3. www.engineeringtoolbox.com/drag-coefficient-d_627.html
4. Magnus effect: http://en.wikipedia.org/wiki/Magnus_effect

REFERENCES

Adhikari, G. R. (1999). Studies on flyrock at limestone quarries. *Rock Mechanics and Rock Engineering*, *32*(4), 291–301. https://doi.org/10.1007/s006030050049

Agrawal, H., & Mishra, A. K. (2021). An analytical approach to measure the probable overlapping of holes due to scattering in initiation system and its effect on blast-induced ground vibration in surface mines. *Mining, Metallurgy and Exploration*, *38*(1). https://doi.org/10.1007/s42461-020-00350-2

Armstrong, L. W. (1994). *The quality of stemming in assessing blasting efficiency*. M.E. Thesis, School of Mines, The University of New South Wales, 145p. https://doi.org/https://doi.org/10.26190/unsworks/5349

Ash, R. L. (1973). *The influence of geological discontinuities on rock blasting*. Ph.D. Thesis (pp. 147–152). University of Minnesota.

Bajpayee, T. S., Rehak, T. R., Mowrey, G. L., & Ingram, D. K. (2004). Blasting injuries in surface mining with emphasis on flyrock and blast area security. *Journal of Safety Research, 35*(1), 47–57. https://doi.org/10.1016/j.jsr.2003.07.003

Bhaskar, A., Baranwal, A. K., Ranjan, P., Jena, T. K., Shkhar, M., & Chakraborty, D. (2019). Application of plastic funnel in blast hole to improve blasting efficiency of opencast coal mine at West Bokaro. In N. Aziz & B. Kininmonth (Ed.), *Proceedings of the 2019 coal operators' conference* (pp. 345–351). University of Wollongong. https://ro.uow.edu.au/coal/752

Blanchier, A. (2013). Quantification of the levels of risk of flyrock. In *Rock fragmentation by blasting, FRAGBLAST 10—proceedings of the 10th international symposium on rock fragmentation by blasting* (pp. 549–553). CRC Press/Balkema. https://doi.org/10.1201/b13759-77

Briggs, L. J. (1959). Effect of spin and speed on the lateral deflection (curve) of a baseball; and the Magnus effect for smooth spheres. *American Journal of Physics, 27*(8), 589–596.

Brinkmann, J. R. (1990). An experimental study of the effects of shock and gas penetration in blasting. *Proceedings of the International Society of Explosives Engineers*, 55–56.

Broadbent, C. D. (1974). Predictable blasting with insitu seismic survey. *Mining Engineering*, 37–41.

Cevizci, H. (2019). Comparison of the efficiency of plaster stemming and drill cuttings stemming by numerical simulation. *Journal of the Southern African Institute of Mining and Metallurgy, 119*(5), 465–470.

Chakraborty, A. K., Raina, A. K., Ramulu, M., Choudhury, P. B., Haldar, A., Sahu, P., & Bandopadhyay, C. (2002). *Development of innovative models for optimisation of blast fragmentation and muck profile applying image analysis technique and subsystems utilisation concept in indian surface coal mining regime*. Ministry of Coal, Govt. of India.

Chernigovskii, A. A. (1985). *Application of directional blasting in mining and civil engineering*. Oxidian Press India Private Ltd.

Chiappetta, R. F., Bauer, A., Dailey, P. J., & Burchell, S. L. (1983). Use of high-speed motion picture photography in blast evaluation and design. In *Proceedings of the annual conference on explosives and blasting technique* (pp. 258–309). International Society of Explosive Engineers.

Choudhary, B. S., & Agrawal, A. (2022). Minimization of blast-induced hazards and efficient utilization of blast energy by implementing a novel stemming plug system for eco-friendly blasting in open pit mines. *Natural Resources Research, 31*(6), 3393–3410. https://doi.org/10.1007/s11053-022-10126-8

Choudhary, B. S., & Rai, P. (2013). Stemming plug and its effect on fragmentation and muckpile shape parameters. *International Journal of Mining and Mineral Engineering, 4*(4), 296–311. https://doi.org/10.1504/IJMME.2013.056854

Correa, C. E., & Navarrete, M. F. (2004). Assessment of stemming plug plastic elements to improve blasting gases confinement in escondida. *Australasian Institute of Mining and Metallurgy Publication Series*, 96–100.

Coulibaly, D. A. (1982). *Effects of stemming on open pit bench blasting*. University of Nevada, Reno. http://scholarworks.unr.edu:8080/bitstream/handle/11714/1209/Mackay133_Coulibaly.pdf?sequence=1&isAllowed=y

Da Gama, C. D. (1984). Microcomputer simulation of rock blasting to predict fragmentation. In *The 25th US symposium on rock mechanics (USRMS)* (pp. 1018–1030). American Rock Mechanics Association.

Dick, R. D., Weaver, T. A., & Fourney, W. L. (1990). An alternative to cube-root scaling in crater analysis. In *Proceedings of the 3rd international symposium on rock fragmentation by blasting, FRAGBLAST 3* (pp. 167–170). AusIMM.

Dobrilović, M., Ester, Z., & Janković, B. (2005). Measurement in blast hole stem and influence of steming material on blasting quality. *Rudarsko Geolosko Naftni Zbornik, 17*, 47–53.

Dowding, C. H. (1985). *Blast vibration monitoring and control* (1st ed.). Prentice-Hall, Inc.

Eloranta, J. (1994, January). Stemming selection for large-diameter blastholes. In *Proceedings of the conference on explosives and blasting technique* (pp. 255–255). International Society of Explosives Engineers.

Falcao Neves, Paula & Costa e Silva, Matilde & Paneiro, Gustavo & Bernardo, Pedro & Gonçalves, João. (2016). Blasting performance when using articulated air-deck stemming systems with plug. 10.5593/SGEM2016/B12/S03.016.

Favreau, R. F., & Lilly, D. (1986). Use of computer blast simulations to evaluate the effect of angled holes in cast blasting. In Y. J. Wang, R. L. Grayson, & R. Sanford (Eds.), *Use of computers in the coal industry* (pp. 143–152). CRC Press, Taylor & Francis Group. https://doi.org/10.1201/9781003079262-23

Gustafsson, R. (1981). *Blasting technique*. Dynamit Nobel Wien.

Hartmann, I., Howarth, H. C., & Nagy, J. (1950). *Experiments on safety of incombustible plugs for stemming explosives* (Vol. 4686). US Department of the Interior, Bureau of Mines.

Higgins, A., Loiseau, J., & Mi, X. C. (2018). Detonation velocity/diameter relation in gelled explosive with inert inclusions. *AIP Conference Proceedings, 1979*. https://doi.org/10.1063/1.5044891

Hustrulid, W. (1999). *Blasting principles for open pit mining: Volume 1—General design concepts*. Balkema.

Jimeno, C. L., Jimeno, E. L., & Carcedo, F. J. A. (1995). *Drilling and blasting of rocks*. Balkema. https://doi.org/10.1201/9781315141435-20

Jimeno, E. L., & Muniz, E. (1987). A new method for design of bench blasting. In *2nd international symposium on rock fragmentation by blasting* (pp. 302–307). Society for Experimental Mechanics.

Joslin, R. D. (1982). *Effect of stemming variations in open pit blasting- a laboratory study*. University of Nevada, Reno.

Kim, H.-S., & Lee, S.-E. (2011). Stemming effect of the crushed granite sand as fine aggregate at the mortar blasting test. *Tunnel and Underground Space, 21*(4), 320–327.

Kim, M. (2022). Projectile app. *MATLAB Central File Exchange*. www.mathworks.com/matlabcentral/fileexchange/52222-projectile-app

King, B. M., Just, G. D., & McKenzie, C. K. (1988). Improved evaluation concepts in blast design. *Transactions of the Institution of Mining and Metallurgy, Section A: Mining Technology, 97*, 173–181.

Kinoshita, N., Kubota, S., Saburi, T., Ogata, Y., & Miyake, A. (2011). Influence of charge diameter on detonation velocity and reaction zone of ANFO explosive contained in a steel tube. *Science and Technology of Energetic Materials, 72*(1–2), 21–25.

Ko, Y., Shin, C., Jeong, Y., & Cho, S. (2022). Blast hole pressure measurement and a full-scale blasting experiment in hard rock quarry mine using shock-reactive stemming materials. *Applied Sciences, 12*(17), 8629. https://doi.org/https://doi.org/10.3390/app12178629

Konya, C. J., & Davis, J. (1978). The effects of stemming consist on retention in blastholes. *Proceedings of the International Society of Explosives Engineers, 10*.

Konya, C. J., & Konya, A. (2018). Effect of hole stemming practices on energy efficiency of comminution. *Green Energy and Technology*, 31–53. https://doi.org/10.1007/978-3-319-54199-0_3

Langefors, U., & Kihlstrom, B. (1978). *The modern techniques of rock blasting*. John Wiley and Sons, Inc.

Leighton, J. C. (1982). *Development of a correlation between rotary drill performance and controlled blasting powder factors*. University of British Columbia.

Livingston, C. W. (1956). Fundamentals of rock failure. *Quarterly of the Colorado School of Mines*, *51*(3), 1–11.

Lubbe, C., & Frye, R. (1998). Function analysis of stemming devices. *International Society of Explosives Engineers*, *24*.

Lundborg, N. (1974). *The hazards of flyrock in rock blasting* (SweDeFo Reports DS1974). Swedish Detonic Research Foundation.

Lundborg, N., Persson, A., Ladegaard-Pedersen, A., & Holmberg, R. (1975). Keeping the lid on flyrock in open-pit blasting. *Engineering and Mining Journal*, *176*(5), 95–100. https://doi.org/10.1016/0148-9062(75)91215-2

Magnus, G. (1853). Ueber die Abweichung der Geschosse, und: Ueber eine auffallende Erscheinung bei rotirenden Körpern. *Annalen der Physik*, *164*(1). https://doi.org/10.1002/andp.18531640102

Maram, D., & Uttarwar, M. D. (2020). Application of stemming systems and materials in underground excavations—a critical review. *International Journal of Innovative Research in Science, Engineering and Technology*, *9*(2), 14010–14021. https://doi.org/10.15680/IJIRSET.2020.0902011

McCleskey, F. (1988). *Drag coefficients for irregular fragments*. Naval Surface Warfare Center.

McHugh, S. (1983). Crack extension caused by internal gas pressure compared with extension caused by tensile stress. *International Journal of Fracture*, *21*(3). https://doi.org/10.1007/BF00963386

McKenzie, C. (2009). Flyrock range and fragment size prediction. In *Proceedings of the 35th annual conference on explosives and blasting technique* (Vol. 2, p. 2). http://docs.isee.org/ISEE/Support/Proceed/General/09GENV2/09v206g.pdf

McNail, C., Schillie, J., & Worsey, P. N. (1999). Pneumatic stemming of horizontal holes with particulate material. *Proceedings of the International Society of Explosives Engineers*, *7*.

Mertuszka, P., Cenian, B., Kramarczyk, B., & Pytel, W. (2018). Influence of explosive charge diameter on the detonation velocity based on Emulinit 7L and 8L Bulk emulsion explosives. *Central European Journal of Energetic Materials*, *15*(2), 351–363. https://doi.org/10.22211/cejem/78090

Miyake, A., Kobayashi, H., Echigoya, H., Kato, K., Kubota, S., Wada, Y., Ogata, Y., & Ogawa, T. (2008). Diameter effect on detonation velocity of ammonium nitrate and activated carbon mixtures. *Materials Science Forum*, *566*, 101–106. https://doi.org/10.4028/www.scientific.net/msf.566.101

Mohamad, E. T., Armaghani, D. J., Hajihassani, M., Faizi, K., & Marto, A. (2013a). A simulation approach to predict blasting-induced flyrock and size of thrown rocks. *Electronic Journal of Geotechnical Engineering*, *18 B*, 365–374.

Mohamad, E. T., Armaghani, D. J., & Motaghedi, H. (2013b). The effect of geological structure and powder factor in flyrock accident, Masai, Johor, Malaysia. *Electronic Journal of Geotechnical Engineering*, *18 X*, 5661–5672.

Mohamad, E. T., Murlidhar, B. R. M., Armaghani, D. J., Saad, R., & Yi, C. S. (2016). Effect of geological structure and blasting practice in fly rock accident at Johor, Malaysia. *Jurnal Teknologi*, *78*(8–6), 15–21. https://doi.org/10.11113/jt.v78.9634

Mohamad, E. T., Yi, C. S., Murlidhar, B. R., & Saad, R. (2018). Effect of geological structure on flyrock prediction in construction blasting. *Geotechnical and Geological Engineering*, *36*(4), 2217–2235. https://doi.org/10.1007/s10706-018-0457-3

Mohanty, B., & Singh, V. K. (2012). *Performance of explosives and new developments*. CRC Press.

Mpofu, M., Ngobese, S., Maphalala, B., Roberts, D., & Khan, S. (2021). The influence of stemming practice on ground vibration and air blast. *Journal of the Southern African Institute of Mining and Metallurgy*, *121*(1), 1–10. https://doi.org/10.17159/2411-9717/1204/2021

Muftuoglu, Y. V., Amehmetoglu, A. G., & Karpuz, C. (1991). Correlation of powder factor with physical rock properties and rotary drill performance in Turkish surface coal mines. In *7th ISRM congress*, International Society of Rock Mechanics.

O'Meara, R. (1994). Blasting over 40 feet of toe burden—a case study to outline modern planning techniques. In *Proceedings of the conference on explosives and blasting technique* (pp. 377–387). International Society of Explosives Engineers.

Oates, T. E., & Spiteri, W. (2021). Stemming and best practice in the mining industry: A literature review. *Journal of the Southern African Institute of Mining and Metallurgy*, (121), 415–426. https://doi.org/10.17159/2411-9717/1606/2021

Pal Roy, P. (2005). *Rock blasting: Effects and operations*. CRC Press.

Pal Roy, P., & Dhar, B. B. (1993). Rock fragmentation due to blasting—a scientific survey. *The Indian Mining and Engineering Journal*, *32*(9), 27–32.

Persson, P. A., Holmberg, R., & Lee, J. (1994). *Rock blasting and explosives engineering*. CRC Press. https://doi.org/10.5860/choice.31-5469

Pradhan, G. K. (1996). *Explosives and blasting techniques*. Mintech Publications.

Rad, H. N., Bakhshayeshi, I., Wan Jusoh, W. A., Tahir, M. M., & Foong, L. K. (2020). Prediction of flyrock in mine blasting: A new computational intelligence approach. *Natural Resources Research*, *29*(2), 609–623. https://doi.org/10.1007/s11053-019-09464-x

Rai, P., Ranjan, A. K., & Choudhary, B. S. (2008). Achieving effective fragmentation: A case study on the role of the stemming column on fragmentation. *Quarry Management*, *35*(2), 17.

Raina, A. K., & Choudhury, P. B. (2014). Demystifying the powder factor. In P. Sen & B. S. Choudhary (Eds.), *6th national seminar on surface mining (NSSM)* (pp. 161–167). Department of Mining Engineering, Indian School of Mines.

Raina, A. K., Murthy, V. M. S. R., & Soni, A. K. (2015). Flyrock in surface mine blasting: Understanding the basics to develop a predictive regime. *Current Science*, *108*(4), 660–665. https://doi.org/10.18520/CS/V108/I4/660-665

Ranganath, P. (2022). Projectile motion: Animation of numerical solutions. *MATLAB Central File Exchange*. www.mathworks.com/matlabcentral/fileexchange/47262-projectile-motion-animation-of-numerical-solutions

Rathore, S. S., & Lakshminarayana, V. (2006). Studies on flyrock at limestone quarries. *Journal of Mines, Metals and Fuels*, *54*(6–7), 130–134.

Rehman, A. U., Emad, M. Z., & Khan, M. U. (2019). *Ergonomic selection of stemming plugs for quarry blasting operation* (p. 5). Southern African Institute of Mining and Metallurgy.

Rehman, A. U., Emad, M. Z., & Khan, M. U. (2021). Improving the environmental and economic aspects of blasting in surface mining by using stemming plugs. *Journal of the Southern African Institute of Mining and Metallurgy*, *121*(7), 369–377.

Rustan, A., Cunningham, C. V. B., Fourney, W., & Spathis, A. (2011). Mining and rock construction technology desk reference. In A. Rustan (Ed.), *Mining and rock construction technology desk reference*. CRC Press. https://doi.org/10.1201/b10543

Sazid, M., Saharan, M. R., & Singh, T. N. (2011, October). Effective explosive energy utilization for engineering blasting–initial results of an inventive stemming plug, SPARSH. In *12th ISRM congress* (pp. 505–506). International Society of Rock Mechanics.

Schwengler, B., Moncrieff, J., & Bellairs, P. (2007). Reduction of the blast exclusion zone at the black star open cut mine. *Australasian Institute of Mining and Metallurgy Publication Series*, 51–58.

Shahrin, M. I., Abdullah, R. A., Jeon, S., Jeon, B., Sa'ari, R., & Alel, M. N. A. (2019). Numerical simulation on the effect of stemming to burden ratio on rock fragmentation by blasting. In *5th ISRM young scholars' symposium on rock mechanics and international symposium on rock engineering for innovative future, YSRM 2019* (pp. 909–914). International Society of Rock Mechanics.

Sharma, S. K., & Rai, P. (2015). Investigation of crushed aggregate as stemming material in bench blasting: A case study. *Geotechnical and Geological Engineering*, *33*(6), 1449–1463. https://doi.org/10.1007/s10706-015-9911-7

Silva, J., Li, L., & Gernand, J. M. (2018). Reliability analysis for mine blast performance based on delay type and firing time. *International Journal of Mining Science and Technology*, *28*(2), 195–204. https://doi.org/10.1016/j.ijmst.2017.07.004

Snelling, W. O., & Hall, C. (1928). *The effect of stemming on the efficiency of explosives* (Vol. 17). US Government Printing Office.

Stojadinović, S., Lilić, N., Pantović, R., Žikić, M., Denić, M., Čokorilo, V., Svrkota, I., & Petrović, D. (2013). A new model for determining flyrock drag coefficient. *International Journal of Rock Mechanics and Mining Sciences*, *62*, 68–73. https://doi.org/10.1016/j.ijrmms.2013.04.002

Tang, Z., Zhieheng, Z., Wen, L. U., & others. (1998). The influence of stemming and stemming methods in blasting boreholes to blasting effect. *Journal of Southwest Institute of Technology*, *13*(2), 64–66.

Thornton, D., Kanchibotla, S. S., & Brunton, I. (2002). Modelling the impact of rockmass and blast design variation on blast fragmentation. *Fragblast*, *6*(2), 169–188. https://doi.org/10.1076/frag.6.2.169.8663

Tripathy, G. R., Shirke, R. R., & Gupta, I. D. (1999). On the importance of seismic wave velocity in rock blasting. *Journal of Rock Mechanics and Tunnelling Technology*, *5*, 97–108.

Trivedi, R., Singh, T. N., & Gupta, N. (2015). Prediction of blast-induced flyrock in opencast mines using ANN and ANFIS. *Geotechnical and Geological Engineering*, *33*(4), 875–891. https://doi.org/10.1007/s10706-015-9869-5

Vasconcelos, H., Marinho, L. F., Lusk, B., Calnan, J., & Sainato, P. (2016). *Application of non-newtonian fluid mixture for stemming performance optimization* (p. 10p.). International Society of Explosives Engineers.

Wilkins, M. D. (1997). *Stemming technique for loading angled holes charged with ANFO*. University of Missouri-Rolla. http://merlin.lib.umsystem.edu/record=b3968565~S5

Worsey, P. (1993). Stemming ejection comparison of conventional stemming and stemming incorporating blast control plugs for increasing explosion energy use. *Explosives Engineering*, (April), 12–19.

Zhang, Z. X., Hou, D. F., Guo, Z., & He, Z. (2020). Laboratory experiment of stemming impact on rock fragmentation by a high explosive. *Tunnelling and Underground Space Technology*, *97*. https://doi.org/10.1016/j.tust.2019.103257

4 Prediction of Flyrock Distance

The definition of flyrock as specified by ISRM (Rustan et al., 2011), that was introduced along with other definitions in Sections 2.1.1 and 2.1.2, is briefly introduced here:

> *Any rock fragments thrown unpredictable from a blasting site. Flyrock may develop in the following situations: insufficient stemming, too small a burden because of overbreak from the preceding or proceeding blast, planes of weakness in rock which reduce the resistance to blasting, and finally existence of loose rock fragments on top of the bench.*

The use of the term "*unpredictable*" in the aforementioned definition of flyrock has some important connotations:

1. The flyrock cannot be predicted. Accordingly, by this definition, whether flyrock will occur in a particular condition cannot be predicted. Practically, there are very little studies on the occurrence of flyrock. However, limited studies have defined the probability of flyrock and hence contradict the aforementioned predicament of unpredictability. If this condition is literally accepted, then there are severe issues with the predictability of flyrock itself.
2. The distance of the flyrock cannot be predicted: there have been significant studies that contradict the definition, if taken on its face value. Flyrock distance prediction has been logged by several authors.
3. The efforts to predict flyrock are not in tune with the phenomenon.
4. The causes of flyrock are well-understood but are not properly incorporated in the flyrock distance prediction models.

Irrespective of the aforesaid statements, specifically chosen words for the definition delineate the conditions that result in flyrock, viz. insufficient stemming (length), too small a burden because of overbreak from the previous blasts, planes of weakness, and lose fragments on top. These conditions, along with others, described in detail in Chapter 3, are not blast design considerations. These are at the most aberrations arising out of several reasons brough out earlier. However, most of the equations, proposed recently for prediction of flyrock distance, use prediction through blast design, while ignoring the rockmass conditions and the role of explosive and delay malfunctions. Yet, a question arises, whether there should be efforts to predict the flyrock distance. McKenzie (2009) provides a rationale for such prediction while stating that flyrock distance prediction is important as it assists in determination of

DOI: 10.1201/9781003327653-4

appropriate clearance distances, helps in determination of best charge configurations while blasting very near to sensitive structures, and helps to evaluate the risk due to flyrock, particularly to the objects that are very near to the blast area. So, the prediction of flyrock distance continues to be an interesting topic ever as also evidenced by the increase in publications on the subject after 2014. There have been attempts to relate the flyrock with the drill diameter and more recently with the design variables. But the models for predicting flyrock distance are still in their infancy and is supported by the following facts.

1. The models in vogue mention "flyrock prediction" which is ambiguous, as there are two components in such prediction, viz. (a) flyrock occurrence that means whether flyrock will occur in a blast or not, and is a probabilistic proposition, (b) flyrock distance prediction that has been and can be treated through empirical or mathematical formulations. Also, there are other probabilities associated with the distance of travel which are defined by change in distance from a blast to the object of concern. When risk due to flyrock is mentioned, it also means defining all the associated probabilities of the event (flyrock occurrence) and the consequences thereof.
2. There are lacunae in flyrock distance prediction models pertaining to the anomalies in objective definition of the solutions sought and the equations derived.
3. Flyrock distance prediction has several aspects and falls in different domains of phenomenon of ejection from blast face, post-ejection interaction in air and rebound, with lot of variables defining its occurrence and travel distance, as elaborated in Chapter 3.
4. While other such unwanted effects like ground vibration, air overpressure, fumes, dust, etc. are regular features, flyrock is uncertain and needs a special treatment. A question arises as to whether the modelling of flyrock distance amounts to predicting the outliers. If this is the case, what are the measures that can be incorporated to account for the rock, explosive, design, and delay malfunction while predicting the flyrock distance?
5. The stringent stipulations deter reporting of flyrock and is rarely reported (Davies, 1995) and hence lack of valid data. This issue will be discussed further in Chapter 5.

In addition, experimentation with flyrock distance modelling is also a difficult proposition because of the following reasons:

1. Flyrock is difficult to simulate in field conditions owing to multitude of modelling constraints.
2. Flyrock cannot be generated and one has to wait for a chance while taking vulnerability of the "Objects of Concern" (Raina et al., 2015b) into account.
3. Controlled studies in mines are important to have a tangible solution to the flyrock occurrence. This means having large data from different drill, blast, and rockmass conditions. Mines, however, refrain from such studies as reporting of flyrock event involves penalties.

However, it is beyond doubt that estimation of flyrock distance is fundamental to any flyrock prediction model that finally leads to definition of blast danger zone (BDZ), which in turn is the focus of the regulatory agencies. Any scientific investigation on flyrock should thus address all the issues raised earlier. Further, enumerating and understanding all possible causes of the flyrock for the sake of its prediction and control is thus important.

4.1 THEORETICAL CONCEPTS

A careful examination of concepts of blasting breakage (Sanchidrián et al., 2007; Udy & Lownds, 1991) reveals that strain energy in burden at escape that contributes to kinetic energy amounts for more than 20% of the total available energy (Table 4.1). Around 18–28% of the energy is logged as lost energy that may be through stemming column due to stemming ejection and variation in burden distance. A small fraction of this energy, e.g. 1% (Berta, 1990), can generate flyrock. The total pressure that will act on a rock fragment that translates to flyrock will be defined by the blasthole pressure and its area under impact, while assuming that there will be no other forces acting on it to retard its motion.

The basic modes of flyrock are documented by several authors, e.g. Little (2007), Persson et al. (1994), and Richards and Moore (2004), that have been further reproduced by multiple authors in recent times and include the following:

1. *Face Burst:* The mechanism involves release of high-pressure explosive gases through the front face which can be a cause of severe flyrock. This generally occurs when there are planes of weakness in burden or actual burden is much less than the designed burden.
2. *Cratering:* When explosive pressures are released from top of the face which can result in formation of crater and significant flyrock.
3. *Rifling:* The condition arises, if the blasthole is highly confined and unable to break and move the burden. The high-pressure explosive gases vent

TABLE 4.1
Explosive Energy Partitioning during Blasting as Reported by Different Authors

Authors	Percentage in Energy Partitioning							
	1	2	3	4	5	6	7	8
Udy and Lownds (1991)	22	14	24	22	18			
Ouchterlony (2004)	10–20		0.1–2				5–10	20–40
Sanchidrián and Ouchterlony (2017)	3–21	9	2–6	1–3			9	
Calnan (2015)	$25^{\#1} + 5^{\#2}$				28	13		

Note: 1—Kinetic energy, 2—Strain energy around blasthole, 3—Fragmentation and heave energy, 4—Strain energy in burden at escape, 5—Lost energy, 6—Borehole chambering (%), 7—Seismic energy, 8—Shock Energy, #1—Rotational kinetic energy, #2—Translational kinetic energy.

through the blasthole that acts like a gun barrel and hence the name riffling. Improper and no stemming may also produce such effect.

The basis of separating face burst from cratering seems to be weak as both the mechanisms pertain to cratering, as initially indicated by Holmberg (1978), except that the direction of cratering is different, i.e. horizontal and vertical directions, respectively. In the case of extreme confinement and no breakage of rock, where a borehole acts as a barrel, the rifling mechanism comes into play. However, in practice, this is a very rare phenomenon. Hence, flyrock can be considered to occur in cratering conditions, irrespective of its direction. An analogy to the barrel can nonetheless be drawn where the orientation of the barrel mimics the least resistant path in the blast bench and hence the flyrock direction.

Further, the peak strains generated during explosion (Duvall & Atchison, 1957; Duvall & Petkof, 1959) attenuates rapidly due to fracturing and approximates to 30–40% of the peak strain. This marks the onset of secondary strain pulse and beginning of secondary cracking that grow at sustained rate (Khanukayev, 1991) and is considered to be the start of the heaving process. However, visual analysis of videos reveals that flyrock ejects quite ahead of any mass movement in the broken rock due to the presence of least resistance paths, of any nature, for the explosive gases. The results indicate that the actual pressures are quite high that propel flyrock at high velocities in comparison to that of the heaving material. The proposition is again akin to the mechanism of cratering for flyrock to occur.

There have been several attempts to predict the flyrock distance using fundamental concepts. Such work(s) include Ladegaard-Pedersen and Persson (1973), Lundborg (1974), Lundborg et al. (1975), and Holmberg and Persson (1976) wherein flyrock distance prediction and its associated probabilities have been defined. Roth (1979) attempted to predict the flyrock distance using ballistic equations. Later attempts by Raina et al. (2006a), Raina et al. (2011), and Stojadinović et al. (2011, 2016) detailed the predictive regimes of flyrock by dwelling on its several aspects.

There are several model types of flyrock distance prediction that can be classified into the following:

1. Mathematical models
2. Empirical and semi-empirical models
3. Numerical models
4. Intelligent prediction techniques based on artificial neural network methods

Most of the models that have been developed for flyrock distance prediction have been compiled in this chapter, and a few classical ones focusing on the fundamentals have been elaborated further.

4.2 MATHEMATICAL MODELS

Lemesh and Pozdnyakov (1972) conducted experiments to calculate rock displacement during blasting and monitored the movement of rockmass using three high-speed cameras in benches of 15–18 m, in blasthole diameter of 200 mm, and rock hardness of 1–13 (Protod'ykanov Index). The starting time, direction, magnitude of visible movement, and other features of action of change were measured in the experiments with the help of markers placed at 50–100 m from the blasts. They observed the following:

1. There was no movement in rock during times varying between 15 and 40 ms after detonation.
2. The rockmass moved at a velocity of 15–30 m/s.
3. The movement in the outer end of the stemming recorded is 4–8 ms after detonation and flyrock has a velocity that is 10–30 times that of the muck.

Finally, Lemesh and Pozdnyakov (1972) devised Equation 4.1 to designate the movement of the rock in relationship with the initial velocity and time of different events in blasting.

$$R_T = v_0 \left(t_0 - t_c\right) \tag{4.1}$$

where R_T is the throw of the rock in m, v_0 is the initial velocity of rock in m/s, t_0 is the time from the detonation of charge in s, and t_c is the time of onset of motion of rock in s.

Lundborg (1974) developed theoretical relations between flyrock distance and blasthole diameter considering the basic physics and impact. The work was focused on determination of maximum possible distance of flyrock from crater blast tests. While calculating the force applied on a flyrock fragment and applying equation of ballistic trajectories, Lundborg (1974) derived the flyrock distance relationship (Equation 4.2) and fragment size (Equation 4.3) by considering that the product of the initial velocity, size, and density of the flyrock ($v_0 \times k_x \times \rho_r$) depends upon the diameter of the blasthole.

$$R_f = 260 d^{\frac{2}{3}} \tag{4.2}$$

$$k_x = 0.1 \times d^{\frac{2}{3}} \tag{4.3}$$

where R_f is the flyrock distance in m, ρ_r is the density of rock in kg/m^3, d is the diameter of the drill hole in inches, and k_x is the size of rock fragment in m travelling maximum distance.

However, in actual mining conditions, the blasthole diameter is constant and hence the predicted flyrock distance assumes similar values irrespective of blast conditions. The size of the flyrock though established through this investigation (Equation 4.3) also yields a constant value in a constant diameter blasthole and thus cannot be related to the in situ block size of the rockmass and the blast design variables.

Further, Lundborg (1974) found that in normal and well-planned bench blasting, the throw is much less than that calculated with Equation 4.2, and proposed a coefficient of ~40 instead of 260 for such blasts in Equation 4.2 and as such, different distances (maximum) of flyrock were predicted for different blasthole diameters. The main constraint with this prediction is that the study was based on hard rock (granite) and the data generated was not of varied type. Hustrulid (1999) commented that "*the equations of Lundborg (1974) provide an approach, although the basis for them is weak and the range of 'hole diameters' limited.*" Also, the effect of the in-hole density of the explosive was not considered in the model that has a direct relationship with blasthole pressure (Cunningham, 2006), that in turn, has a significant bearing on the flyrock distance. In addition, the effect of blast design variables and rock type were not considered. The model, however, forms the basis for later studies in flyrock distance prediction and can be used to estimate the maximum flyrock distance in extreme blast conditions.

Lundborg et al. (1975) calculated the maximum throw distance from field tests conducted in quarries and surface mines by taking into account initial impulse energy and air resistance. He predicted that for 75–250 mm blasthole diameter, the maximum flyrock distance could be in the range of 500–1000 m. He observed from the study that throw and exit velocity are proportional to specific charge (kg/m^3) and approximates to the form of Equation 4.4.

$$R_T = 137q - 25.5,\ q > 0.1 \tag{4.4}$$

where R_T is the throw in m and q is the specific charge in kg/m^3.

The prediction through Equation 4.4 may be limited to the throw only as in general blasts, the maximum displacement of material that can be predicted through the equation may at the most range up to 100 m only. The concept of "q" in relation to blast outcome is however debatable (King et al., 1988).

Roth (1979) developed a relation for obtaining the flyrock distance using ballistic trajectory equation to estimate the exit or launch velocity of flyrock as given in Equation 4.5, while incorporating Gurney's equation (brief explanation provided later in the text).

$$\mathrm{v}_0 = 2E^{0.5} f\left(\frac{q_l}{m_l}\right) \tag{4.5}$$

where $(2E)^{0.5}$ is Gurney's constant, a function of explosive, v_0 is the exit or initial velocity of fragment in m/s, q_l is the linear charge concentration in kg/m, and m_l is the total mass of material per unit of length.

The equation has been modified for flyrock coming from vertical faces (Equation 4.6):

$$v_0 = (2E')^{0.5}\left(\frac{q_l}{m_l}\right) \tag{4.6}$$

where $(2E')^{0.5}$is smaller than $(2E)^{0.5}$ as the direction of detonation is tangent to the rock and v_0 is the initial velocity of the flyrock in m/s.

Roth (1979) suggested a value of $\frac{c_d}{3}$ for $(2E')^{0.5}$ for different types of explosives. For ammonium nitrate fuel oil explosive, the value of $2E'$ has been taken as 0.44. Considering the energy losses, Equation 4.6 transforms into Equation 4.7,

$$v_0 = (2E')^{0.5} \frac{q_l}{m_l} \frac{1 - [k_1 W_s + k_2 W_j]}{E'} - 2k_3 W_r \tag{4.7}$$

where v_0 is the initial velocity of flyrock in m/s, c_d is the velocity of detonation of the explosive (m/s), W_s is the seismic energy generated per unit weight of explosive (J), W_j is the energy required to crush unit weight of rock (J), W_r is the energy absorbed to fragment of a unit weight of rock in J, and k_1, k_2, and k_3 are proportionality constants.

For estimating the travel distance, Roth (1979) has used simple kinematic equation (Equation 4.8):

$$R_f = v_0^2 \frac{\sin 2\theta}{g} \tag{4.8}$$

where R_f is the flyrock distance in m, v_0 is the initial velocity of fragment in m/s, θ is the angle of flyrock projection in degrees, and g is the acceleration due to gravity in m/s^2.

The approach of Roth (1979), however, assumes that there is no impact of post-release effects like air drag during the flight of fragments, thereby overestimating the flyrock distance.

Gurney's Constant: A Note

Gurney (1943) developed a model to describe the expansion of a metal cylinder driven by the detonation of an explosive. The model closely predicts the initial velocity of the fragments produced by the breakup of the cylinder. He assumed the partitioning of energy between a metal cylinder and gases driving it and a linear velocity gradient in the expanding gases. A simple relationship as given in Equation 4.9 is as follows:

$$\frac{v_0}{(2E)^{0.5}} = \left[\frac{m}{Q} + \frac{1}{2}\right] - \frac{1}{2} \tag{4.9}$$

where v_0 is the initial velocity of metal, m is the mass of metal, Q is the mass of explosive charge, and $2E$ is a constant called Gurney's velocity coefficient, which is specific to a particular explosive.

The equation assumed different forms for different explosives. However, a simple method is given in Equation 4.10 to calculate the constant:

$$(2E)^{0.5} = \frac{c_d}{2.97} \tag{4.10}$$

where c_d is the velocity of detonation of explosive.

Chiappetta et al. (1983) derived several expressions as given in Equations 4.11–4.16 relevant to determination of flyrock distance and its initial velocity. While calculating the flyrock distance, the air resistance, wind direction, and speed were not considered. Chiappetta et al. (1983) showed that flyrock, usually but not always, travels less than 90 m and rises only a few tens of metres at maximum. The relationships for height and distance of flyrock developed by them are given in Equations 4.11 and 4.12:

$$R_h = v_0 \; 2\sin\frac{2\theta}{g} \tag{4.11}$$

$$R_f = \mathrm{v}_e cos\frac{\mathrm{v}_e \sin\theta + 2\mathrm{v}_e \sin\theta + 2gH}{g} \tag{4.12}$$

where R_h is the distance travelled by the flyrock along a horizontal line at the original elevation of the rock on the face in m, R_f is the total distance travelled by a fragment ejected from the blast accounting for its height above the pit floor in m, v_0 is the initial velocity of flyrock in m/s, θ is the launch angle in degrees, and g is the acceleration due to gravity in m/s^2.

The weight of the rock does not enter Equations 4.11 and 4.12 directly. However, for a given applied force, smaller flyrock fragments will have greater acceleration that equals the applied force divided by the mass of the rock. Therefore, one would expect to find smaller fragments at longer distances from the blast, as defined in Equations 4.13 and 4.14:

$$H_1 = \frac{V_i^2 \sin^2\theta}{2g} \tag{4.13}$$

$$H_2 = \frac{V_i^2 \sin^2\theta}{2g} + H \tag{4.14}$$

where H_1 is the maximum height along the distance R_f, and H_2 is the maximum height along the distance R_h.

To calculate the height to which the flyrock fragment was raised at any point on the trajectory, Chiappetta et al. (1983) devised Equation 4.15:

$$H_i = R_f\left[\tan\theta - \frac{0.5gR_h}{v_0^{\,2}\cos\theta}\right] \tag{4.15}$$

where v_0 is initial velocity of the flyrock in m/s, H_i is the height of rise in m at any given horizontal distance measured relative to the original elevation of the fragment, R_h is the horizontal distance along the trajectory in m, and R_f is the total flyrock distance in m.

Finally, the time a piece of flyrock will take to travel a given horizontal distance can be calculated by using Equation 4.16:

$$t = \frac{R_f}{v_e \cos\theta} \tag{4.16}$$

where t is the travel time in seconds and R_f is the flyrock distance in m.

The authors (Chiappetta et al., 1983) are doubtful that computational techniques alone will consistently give accurate result for flyrock velocity considering that rockmass is a non-homogeneous material. The problem with their approach is use of ballistic equations while the flyrock movement is under the influence of air drag, rotation, wind direction, and the wind velocity.

4.3 EMPIRICAL AND SEMI-EMPIRICAL MODELS

Chiappetta et al. (1983) defined the velocity of the flyrock in terms of burden and linear energy of explosive (Equation 4.17):

$$v_0 = K\left[\frac{B}{\sqrt[3]{E_{lq}}}\right]^{-1.17} \tag{4.17}$$

where B is the burden (m), S is the spacing (m), V is the flyrock speed in m/s, E_{lq} is the linear energy of the explosive charge in $\times 10^3$ J/m, K is a coefficient expressing the probability of attaining the estimated speed $P(v)$ and ranges between 14 and 54 with $P(v)$ variation from 50% to 0.01% with intermediate values of 25, 32, and 40.7 for $P(v)$ of 5%, 1%, and 0.1%, respectively (Blanchier, 2013).

Gupta (1990) established an empirical relation among the stemming length, burden, and flyrock distance (Equation 4.18):

$$\frac{l_s}{B} = 155.2R_f^{-1.37} \tag{4.18}$$

where l_s is the length of stemming in m, B is the burden in m, and R_f is the flyrock distance in m.

They observed that the reduction in the distance of flyrock was not appreciable for stemming length to burden ratio greater than 0.6. A small change in the ratio below 0.2 resulted significant increase in flying distance. However, considering the dynamics of the blast, burden is an important factor for horizontal throw, whereas change in stemming length influences vertical throw or cratering. The method though does not include the rock and explosive properties.

Flinchum and Rapp (1993) explained the procedures for recognizing open face and bench blasting by various surveying methods and provided mathematical formula for calculating and designing proper blasthole placement and explosive load configuration. A concept called "Pattern Footage" was developed by them in a USBM study to minimize air-blast and flyrock. The pattern footage is determined by using Equation 4.19:

$$f_p = \left[\frac{d_e}{12}\right]^2 k \quad (4.19)$$

where f_p is the pattern footage, d_e is the explosive diameter in mm, and k is an empirical constant.

By utilizing the pattern footage from Equation 4.19, burden is determined by Equations 4.20 and 4.21:

$$\sqrt{f_p} = Burden \text{ (Square pattern)} \quad (4.20)$$

or

$$\sqrt{f_p} = 0.85 \times Burden \text{ (rectangular pattern)} \quad (4.21)$$

Desired drill pattern spacing is determined by $S = \frac{f_p}{B}$ where toe burden may be greater than the pattern burden. The authors say that the information is not suitable under all conditions. They advised users to make their own tests to determine the safety and suitability of this information for specific purposes.

Workman and Calder (1994) predicted the flyrock travel distance by neglecting the air resistance, wind direction, and speed. They considered that the flyrock usually and not always travel less than 3000 ft. (~900 m) and rises only a few hundred feet at maximum.

Richards and Moore (2004) expanded the work of Workman and Calder (1994) while classifying the three modes of flyrock, viz. "rifling," "face burst," and "cratering," developed a scaled burden method for calculation of exit velocity as defined by Equations 4.22–4.24 for evaluating the possible flyrock distance:

$$\text{Face burst: } R_f = \frac{k^2}{g}\left[\frac{\sqrt{q_l}}{B}\right]^{2.6} \quad (4.22)$$

$$\text{Cratering: } R_f = \frac{k^2}{g}\left[\frac{\sqrt{q_l}}{l_s}\right]^{2.6} \quad (4.23)$$

$$\text{Rifling: } R_f = \frac{k^2}{g}\left[\frac{\sqrt{q_l}}{l_s}\right]^{2.6} \sin 2\alpha_{bi} \quad (4.24)$$

where α_{bi} is blasthole angle in degree, R_f is the flyrock distance in m, q_l is the explosive mass/m in kg/m, B is the burden in m, l_s is the stemming length in m, g is the acceleration due to gravity (m/s^2), and k is a constant.

The horizontal throw (R_f) reached by flyrock to a point "H" metre above the launch site is determined (Richards & Moore, 2004) from Equation 4.25, which is again a derivative of the trajectory equation:

$$R_f = v_0 \cos\theta \left(\frac{v_0\ Sin\theta + \sqrt{(v_0\ Sin\theta)^2 - 2gH}}{g} \right) \tag{4.25}$$

where v_0 is the launch velocity(m/s), θ is the launch angle (from horizontal), g is the gravitational constant, and R_f is the horizontal throw.

The launch velocity can be determined from Equation 4.26:

$$v_0 = k\left(\frac{\sqrt{m}}{S.H.}\right)^{1.3} \sin 2\theta \text{ or } \sqrt{\frac{R_f g}{\sin 2\theta}} \tag{4.26}$$

The maximum height reached by flyrock at the top of its trajectory is

$$H = \frac{v_0^2 Sin^2\theta}{2g}$$

where m is the explosive quantity per delay in kg and k is a constant.

However, the post-release effect on flyrock distance due to air drag, etc. has not been incorporated in the previous equations.

St. George and Gibson (2001) presented a mechanistic Monte Carlo method to predict exit velocity of flyrock while incorporating air drag as given in Equation 4.27. They used the concept of borehole pressure based on half of the *CJ* pressure for calculating the initial velocity of the flyrock.

$$v_e = \frac{3\rho_e c_d^{\,2} \Delta t}{32 k_x \rho_r} \tag{4.27}$$

where ρ_e is density of explosive in $\times 10^{-3}$ kg/m^3, c_d is the velocity of detonation of explosive in m/s, Δt is the length of impulse time in s, ρ_r is the density of rock in $\times 10^{-3}$ kg/m^3, and k_x is the diameter of the flyrock in $\times 10^{-2}$ m.

Little and Blair (2010), however, questioned their assumptions of borehole pressure.

Raina et al. (2006a) devised a factor of safety for flyrock (FoS) as a ratio of the forces that resist flyrock and forces that tend to generate flyrock (Equation 4.28), while assuming the specific charge as constant:

$$FS_H = C_f\left(\frac{B_d \times J_{fr}}{d_e}\right) \tag{4.28}$$

where FS_H is the dimensionless factor of safety for horizontal flyrock, C_f is the correction factor for different factors controlling or contributing to flyrock occurrence,

B_d is the drilled burden in m, J_{fr} is the joint frequency rating (100 × joint spacing with a maximum value of 100), and d_e is the diameter of the explosive in m.

The correction factor C_f is provided as follows (Raina et al., 2006a):

1. Favourable conditions: $C_f = 1.1–1.2$
   i. Decking (bottom, top, middle—solid or air decking that reduce charge concentration)
   ii. Use of non-electrical-detonator and shock tube combination that reduce fly rock due to bottom initiation
   iii. Use of in-hole multiple delay technique that divides charge in several segments in a single hole and that blast at different times
   iv. Stemming methods (nicely tamped, use of stone crusher chips, etc.)
2. Unfavourable conditions:
   i. Choke blast or solid blasting, $C_f = 0.7$
   Weak zones (both in horizontal and vertical direction (if no measures are taken)
   ii. Karst features (presence of cavities in the rockmass, $C_f = 0.5–0.6$
   iii. Weak layers within competent rock, $C_f = 0.67$
   iv. Use of detonating fuse as in-hole initiation system, $C_f = 0.8$

A blast is unsafe for FoS < 0.5, likely unsafe for values between 0.5 and 1.0, safe for values of 1.0–2.0, and highly safe for values >2.0. These values FoS < 0.5 map to $R_f > 40$ m.

McKenzie (2009) developed empirical formulas to predict the maximum flyrock distance, and the particle size achieving the maximum distance for blasts of varying rock density, blasthole diameter, explosive density, and degree of confinement in terms of scaled depth of burial of explosive. McKenzie (2009) based his findings on the previous studies of Lundborg (1974) and Lundborg et al. (1975). The flyrock distance was further defined in terms of hole diameter and shape factor, and size of rock fragment that achieves maximum projection distance in terms of rock density and shape factor while defining a velocity coefficient for flyrock as given in Equations 4.29–4.34. The initial calculation for evaluating flyrock distance is to work out the scaled depth of burial (Equation 4.29).

$$B_{sd} = \frac{l_s + 0.0005 l_c d}{0.00923\left(l_c d^3 \rho_e\right)^{0.333}} \quad (4.29)$$

where B_{sd} is the scale depth of burial, l_c is the charge length expressed as a multiple of hole diameter, with a maximum value of 8 for hole diameter (d) less than 0.1 m and a maximum value of 10 for diameters greater than 0.1 m, l_s is the stemming length in m, d is the drill diameter in $\times 10^{-2}$ m, and ρ_e is the density of the explosive in $\times 10^{-3}$ kg/m^3.

The velocity coefficient K_v is defined by Equations 4.30 and 4.31:

$$v_e = K_v \left[\frac{d}{k_x}\right]\left[\frac{2.6}{\rho_r}\right] \quad (4.30)$$

$$K_v = 0.0728 \times B_{sd}^{-3.251} \quad (4.31)$$

where d is the diameter of blasthole in $\times 10^{-3}$ m, v_e is the exit velocity of flyrock in m/s, ρ_r is the density of the rock in $\times 10^{-3}$ kg/m^3, and k_x is the flyrock diameter in $\times 10^{-3}$ m.

McKenzie (2009) also replaced the correction factor for air drag of Chernigovskii (1985), as given in Equation 4.32:

$$b_d = \frac{L_f}{k_x \rho_r} \quad (4.32)$$

where b_d is a dimensionless correction factor, L_f is the shape factor of the flyrock and assumes a value of 1.1–1.3 for most of the flyrock fragments, k_x is the flyrock diameter in $\times 10^{-3}$ m, and ρ_r is the density of the rock in $\times 10^{-3}$ kg/m^3.

Using the aforementioned assertions, McKenzie (2009) expressed the Lundborg's empirical formula for flyrock distance, shape factor in terms of B_{sd}, and flyrock size for achieving maximum throw in Equations 4.33 and 4.34, respectively.

$$R_f = 11 \times B_{sd}^{-2.167} \left[\frac{d}{L_f} \right]^{0.667} \quad (4.33)$$

$$k_x \left[\frac{\rho_r}{2.6} \right] = 2.82 \times B_{sd}^{-2.167} d^{0.667} L_f^{0.333} \quad (4.34)$$

However, McKenzie (2009) concludes that the particle size defined by Equation 4.34 can only be estimated for shape factors defined by him and not for other ranges.

A comprehensive list of all other flyrock distance prediction models that can be found in the literature is provided in Table 4.2.

A lot of variation in blast design variables and use of terms for flyrock distance prediction exists in models presented in Table 4.2. These present simple to complex measures for such calculations. One major drawback with majority of these models is that they do not include rockmass properties or explosive properties and certain causes of flyrock. Only a few equations use rock properties like density of the rock and rockmass rating (Bieniawski, 1989). Only one model includes role of rock impedance and explosive property. In addition, such models, some with very high coefficients of determination, use just blast design variables to model a phenomenon like flyrock distance that is not a design construct. The number of significant digits of estimated parameters, R^2, and too low RMSE values reported in many models have not been explained in relationship with the measurements of independent and dependent variables.

The examination of the flyrock distance model(s) compilation in Table 4.2 brings out an important fact that the flyrock models proposed have focused more on established properties of explosive and blast design parameters. A rethinking is necessary to incorporate the rockmass variables and devise methods to improve the

TABLE 4.2
Comprehensive List of Flyrock Distance Prediction Models of Empirical Nature as Available in the Published Domain

S. No.	Author(s)	Model	R^2	RMSE
1	Aghajani-Bazzazi et al. (2010)	$R_f = \ln(1.77 + 0.32q - 0.259B - 0.090S)$	0.93	-
2	Rezaei et al. (2011)	$R_f = -43.9 - 37.442B + 30.581S + 2.614l_{bh} - 119.355b_{sd} - 6.4l_s + 0.144Q_{max} - 2.256\rho_r + 303.787q$ q is the specific charge (kg/ton)	0.70	8.3
3	Ghasemi et al. (2012)	$R_f = 6946.547 \left[B^{-0.796} S^{0.783} l_s^{1.994} H_b^{1.649} d^{1.766} \left(\frac{q}{Q_h} \right)^{1.465} \right]$	0.83	6.1
4	Ghasemi et al. (2012)	$R_f = 6946.547 \left[B^{-0.796} S^{0.783} l_s^{1.994} l_{bh}^{1.649} d^{1.766} \left(\frac{q}{Q_{avg}} \right)^{1.465} \right]$ Q_{avg} is the mean charge per blasthole		
5	Monjezi et al. (2012)	$R_f = 121.467 - 0.14B - 0.346S + 0.112Q_{max} + 10.986d + 0.919l_{bh} - 46.451l_s - 34.56b_{sd} + 38.604q + 0.621RMR$ b_{sd} is the specific drilling (m/m^3), RMR is rockmass rating	0.88	0.04
6	Khandelwal and Monjezi (2013)	$R_f = 88.1311 - 2.8214l_{bh} - 0.1134S - 2.8338B + 2.6665l_s + 52.7774q - 4789b_{sd}$	0.44	7.74
7	Trivedi et al. (2014)	$R_f = \frac{10^{5.1} q_l^{0.51} q^{0.14}}{B^{0.93} l_s^{0.64} \sigma_c^{0.75} RQD^{0.93}}$ σ_c is the unconfined compressive strength; RQD is the rock quality designation	0.82	3.10
8	Marto et al. (2014)	$R_f = -11.873l_{bh} - 10.296B + 136.128Q_{max} - 14.218q + 0.282\rho_r + 1.562N_r + 9.665$ N_r is the Schmidt hammer rebound number	0.54	34
9	Raina and Murthy (2016)	$\ln(R_f) = 2.22 + 5.6310^{-8} Z_r + 7.6310^{-4} \rho_{ee} - 1.810^{-3}(BS) + 2.14\frac{l_c}{l_{bh}}$ Z_r is the impedance of rock and ρ_{ee} is the equivalent in-hole explosive density determined as a ratio of explosive to blasthole volume		

S. No.	Author(s)	Model	R^2	RMSE
10	Faradonbeh et al. (2016)	$R_f = 2.22d + 4.076 l_{bh}^{1.06} + 54.617 \frac{S}{B}^{0.422} - 77.838 l_s^{0.131} - 2.814 Q_{max}^{0.584} + 85.864 q^{1.45} - 98.186$ $R_f = \left[cos(d) \left(ln \left(\frac{l_{bh} + Q_{max}}{\frac{S}{B}} \right) \right)^2 \right] + \left[l_{bh} + (((0.57 l_s) + 48.75 (1.14 l_s)) \right] + [\frac{B}{S}(l_{bh} - q) + \left[\arctan(q)(2d - \sqrt[3]{d}) \right]$	0.82	26.2
11	Armaghani et al. (2016)	$R_f = 177.81 - (3.33 l_{bh}) - (2.55S) - (3.49B) - (13.93 l_s) + (0.47q) + (1Q_{max}) - (2.58RMR)$	0.86	23.3
12	Faradonbeh et al. (2016)	$R_f = \left[\frac{(l_s \times q) + Atan(9.27185) + B}{(-3.141784 + l_{bh}) \times B} \right] \times Atan \left[\begin{matrix} B(B - S + l_s)\sqrt{B^2} \,] \times Atan[B - 2S + 2l_{bh} \\ -S^2 \,] \times Ln[q - (S \times 5.194366 \times 8.466339) \end{matrix} \right]$		
13	Dehghani and Shafaghi (2017)	$R_f = \exp(0.394) \left(\frac{Q}{q} \right)^{0.06} (d)^{0.59} (l_{bh})^{0.26} (NB)^{0.4} (S)^{0.39} (B)^{0.51} (l_s)^{-0.32} (b_{sd})^{0.33}$	0.91	
14	Hasanipanah et al. (2017b)	$R_f = (-90.62 l_{bh}) - (7.76S) - (4.31B) + (53.99 l_s) + (0.62q) + (8.38 Q_{max}) + 5.23$	0.77	27
15	Hasanipanah et al. (2018)	$R_f = -8.48 - (2.22 Q_{max}) + (0.39q) + \left(6.47 \frac{S}{B} \right) - \left(125 \frac{l_s}{B} \right) + 61.74 \frac{l_{bh}}{B}) + (13.34 \frac{B}{d} - (1.44RMR)$	0.64	26.4
16	Faradonbeh et al. (2018)	$R_f = \left[(\sin(S^2) - (6.678162 \times sin(d))) \times l_s \right] + \left[(S^2 + S) \times cos(d) \times S \times q \right] + \left[(d + 5.542816B) \times (S - 2.623993 + d - B) \right] + (Sin(l_s^2) - d\ Sin(d)) \times S$		
17	Zhou et al. (2020a)	$R_f = 0.39d + 0.44 l_{bh} + 46.4 \frac{S}{B} - 0.27 l_s + 0.21 Q_{max} + 121.65q - 31.76$	0.82	0.098
18	Ye et al. (2021)	$R_f = \left[((d + (d \times BS)) \times q) + \sin(l_s + d) \right] + \left[l_{bh} \sqrt[3]{2.3060 - \frac{S}{B} \cos(q) \times 1.0002} \right] + [\arctan(Q_{max}) - 4.8343 + (2 l_s \sin(Q_{max}))]$		

(Continued)

TABLE 4.2 (Continued)
Comprehensive List of Flyrock Distance Prediction Models of Empirical Nature as Available in the Published Domain

S. No.	Author(s)	Model	R^2	RMSE
19	Monjezi et al. (2021)	$R_f = -15.683 + \left(0.737\frac{B}{\ln(B)} + \frac{\ln(l_s)}{l_s^2}\right) + \left(0.383S^2\ln(S) + \frac{\ln(q)}{q^2}\right)$	0.83	14.7
20	Jamei et al. (2021)	$R_f = 1003.2 - 169.89S - 14.374B - 242.41l_s + 77.832q + 37.201SB - 101.49Sq + 41.739l_s^2$	0.98	14.9
21	Shakeri et al. (2022)	$R_f = 307.24 + \left(\left(-0.17B^3\right) - \left(\frac{9.45}{q}\right)\right) + \left(\left(-0.44B^3\right) + \left(14.22l_s\right)\right) - \left(\frac{366.86}{l_{bh}^2}\right) + \left(\left(273.57B\ln(B)\right) + \left(-195.3B^{1.5}\right)\right)$	0.89	7.3

Note: B is the burden (m), S is the spacing (m), l_{bh} is the hole depth (m), b_{sd} is the specific drilling (m/m³), l_s is the stemming (m), Q_{max} is the charge per delay, ρ_r is the rock density (g/cm³), and d is the blasthole diameter.

predictability of flyrock distance while assigning significant weightage to the causative factors of flyrock discussed in Chapter 3. It may be noted that just blast design variables cannot explain the variance of flyrock as majority of flyrock occurs from anomalies in rockmass, blast design, and/or explosive, and even human factors as will be seen in Chapter 6.

4.4 COMPARATIVE ANALYSIS OF VARIOUS FLYROCK DISTANCE MODELS

A comparative analysis of the flyrock distance prediction models, that are of classical yet fundamental nature, have been employed to determine their reliability while using measured data of flyrock distance from field blasts (Figure 4.1).

The correlation results in Figure 4.1 indicate that the best values are provided by the equations proposed by McKenzie (2009) with an R^2 of 0.45 and St. George and Gibson (2001) with an R^2 of 0.44. The equation of Lundborg (1974) being specific to drill diameter presents a near uniform range of flyrock distance for different diameters, which results in low R^2. Similarly, the equation of Richards and Moore (2004) shows low level of R^2. The predictability of most of the other models introduced in Table 4.2 is contested by the respective authors themselves, while resorting to the intelligent methods for prediction of flyrock distance. It is anticipated that significant research is needed to have a universal agreement on any one criterion. Such criterion should be simple in nature and yet have high level of predictability and reliability.

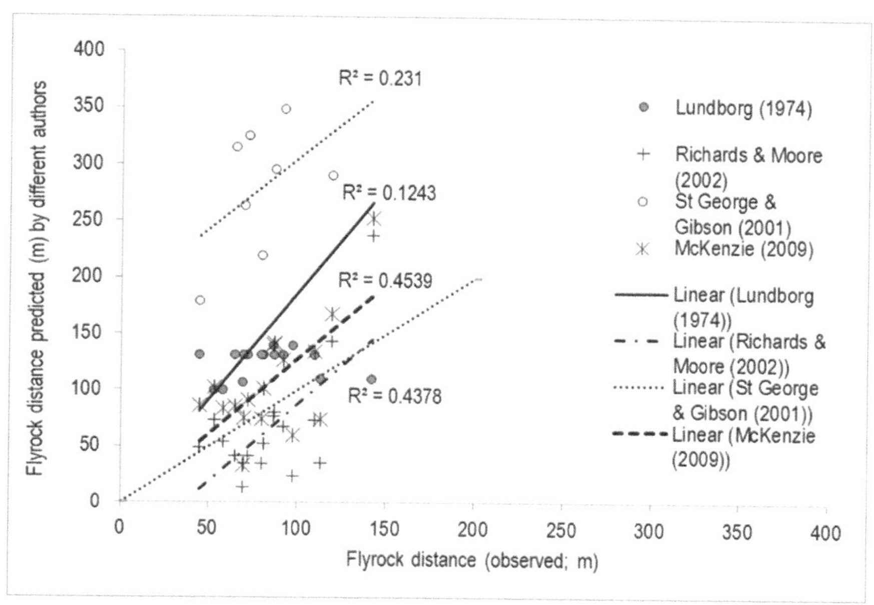

FIGURE 4.1 Comparison of predictability of flyrock distance of cited models.

4.4.1 Summary and Broad Analysis of Data from ANN Reports

The summary data available in more recent publications (Table 4.4) was compiled and analysed further to seek the information of the prediction regimes adopted and the range of data reported. The data reported therein belongs to nine types of rocks/mines with majority from granite, copper, iron, and limestone mines (Figure 4.2). A significant set of data does not mention the type of mines. The data from coal mines is insignificant even though these consume maximum quantity of explosives.

A cross-examination of the data reported reveals analysis of a huge database of 5392 blasts. Nonetheless, major data belong to two mines, e.g. 18 studies from Ulu Tirram Mine with 1727 data points and 5 studies from Sungun Mine with 1165 data points of flyrock (Figure 4.3). It is not clear from such studies whether same set of data was used by different authors for independent studies, or every study represents a new data set. In such a case, it is difficult to ascertain the exact database size. The database, however, represents a wide range of blasthole diameters ranging from 0.05 m to 0.17 m with a mode of around 0.1 m (Figure 4.4). A summary statistic of such database is presented in Table 4.3.

Table 4.3 gives an idea about the extent of variance of data covered that is related to flyrock studies. The data on several aspects, e.g. the number of blasts monitored and the number of blasts that produced flyrock, presence of rock/explosive anomalies, and other non-design variations are not part of such database. This leaves a

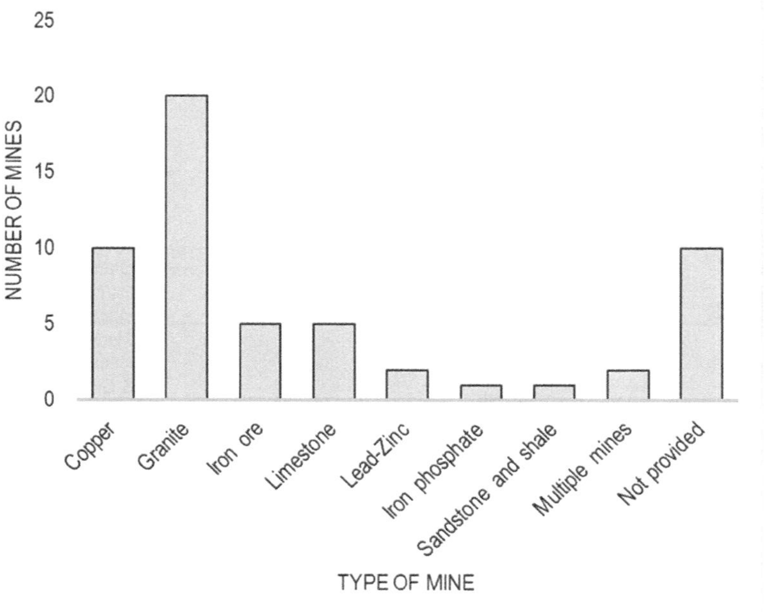

FIGURE 4.2 A number of different types of mines investigated for flyrock in the recent literature.

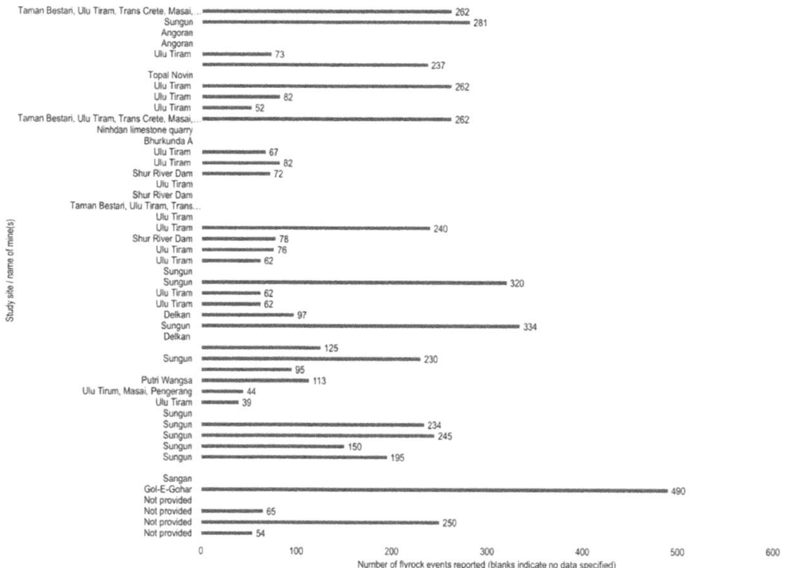

FIGURE 4.3 Number of data sets of flyrock events reported by various authors in recent publications.

significant gap in reliability of the reported results as the probability of flyrock occurrence and nature of flyrock is quite important for such studies.

A correlation between the design variables reported and flyrock distance (minimum, maximum, and average) given in Figure 4.5 reveals that there is significant departure from the expected behaviour and rationale of such derivations. The question is however left for readers to judge.

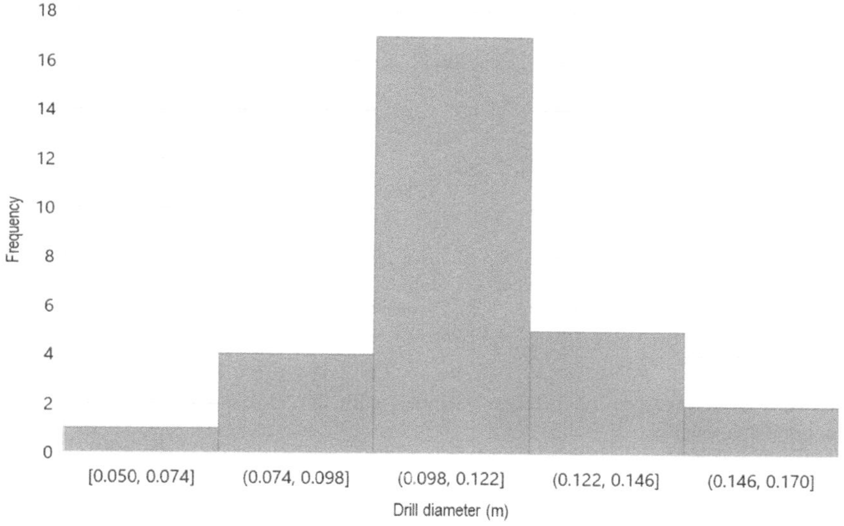

FIGURE 4.4 Blasthole diameters reported in recent publications.

TABLE 4.3
Summary of the Data as Reported in Recent Publications (Count Indicates the Number of Studies Reporting the Use of the Corresponding Variable)

Variable/Factor	Count	Min	Max	Average
Drill Diameter	**29**			
Blasthole depth (m)	14	3	30	14.9
Bench height (m)	18	3	29	12.6
Burden (m)	36	0.63	6.5	3.0
Spacing (m)	15	1.5	6.5	3.6
Stemming length (m)	45	0.5	10	3.0
Specific charge (kg/m^3)	37	0.01	1.72	0.61
Rock density (kg/m^3)	4	2200	4400	Very little data
Flyrock distance (m)	47	10	405	159

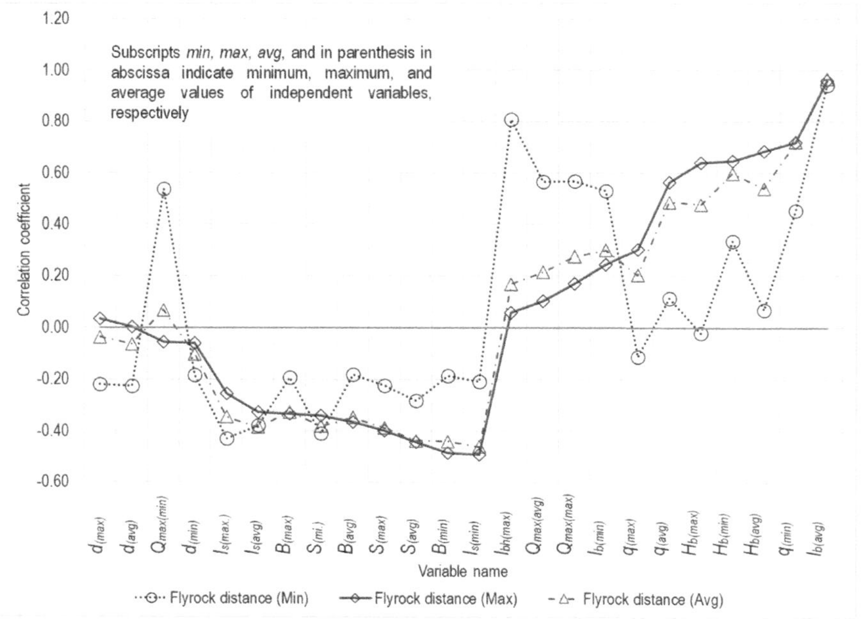

FIGURE 4.5 Correlation of different variables with flyrock distance from data obtained from published papers.

4.5 NUMERICAL CONCEPTS

There are very few numerical solutions to flyrock occurrence possibly because of lot of unknowns and lack of knowledge on boundary conditions for simulation of flyrock. Raina et al. (2006a, 2006b) modelled flyrock using PFC2D (Itasca) using the following details:

> Radius of the balls 0.25 m; density of the balls 2500 kg/m^3; coefficient of friction 0.7; shear and normal stiffness, 1 $\times 10^{-8}$ MPa/m, bond strength, shear 1 $\times 10^{-5}$ MPa, normal 1 $\times 10^{-6}$ MPa and damping coefficient 0.25. Rockmass of massive nature, vertical, horizontal, and both joints together were modelled. Fully coupled holes were assumed in a blasthole diameter of 0.25 m with a bench height of 10 m for simulation. The burden of the blastholes was varied from 4.0 m to 6.0 m over an interval of 1.0 m. Two conditions of initiation viz. bottom hole and top hole (below stemming zone) were also simulated.
>
> Initially the model showing a rock bench is brought to an equilibrium condition. Then the explosion process is created by a sudden and rapid expansion of particles in the blast hole for a very small period to create fractures. The expanded particles are eliminated beyond this period to indicate that the gas flows to the atmosphere and ceases to exert any more pressure. The effects of explosion induced shock waves and the gas expansion are incorporated by running the model for nearly 5000 cycles with one (1) millisecond time step.

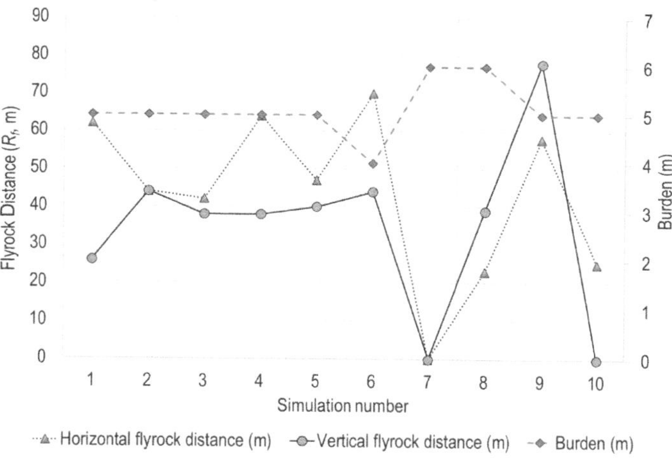

FIGURE 4.6 Results of numerical simulation of flyrock using PFC2D.

The heaving action and the rock fragments displacements, which is comparatively a slower process, is brought to a stable equilibrium condition within this period. The results of the simulations are provided in Figure 4.6.

The following conclusions were drawn from the study:

i. Vertical joint sets contribute to flyrock more in horizontal direction and horizontal joint sets contribute more to the vertical upward direction (simulations 1 and 2). It means that the slabbing effect dominates the occurrence of the flyrock. It also indicates that the maximum flyrock occurs from the front free face in the case of vertical joints and that from the top of the face in the case of horizontal joints.
ii. The increase in the joint frequency may help in reducing the flyrock distance as is evident from cases 1, 2, and 3. The reason may be attributed to the fact that the explosive gas pressure is released easily through various joints, leading to decrease in gas pressure, when there are more joint sets. Hence, greater flyrock is expected in massive formation and vice versa.
iii. The dip of the sheared plane is responsible for deciding the direction of maximum flyrock and its distance (simulations 4 and 5). The sheared or weak planes dipping towards the free face helps to orient the flyrock in downward direction. However, dipping against the free face helps in upward direction of the flyrock. In the latter case, the flyrock distance is greater.
iv. Lesser burden contributes to more horizontal flyrock towards the front of free face (simulation 6), and less stemming results in more vertical flyrock (simulation 9).
v. Proper explosives strength should be selected to have good breakage and to reduce flyrock (simulations 7 and 8) is apparent. The horizontal flyrock

towards the front of the free face is less and nominal than that in the backward direction.

vi. Top "air-decking" shows less flyrock than bottom "air-decking" with restricted horizontal and top side flyrock (simulation 9 and 10). The bottom air decking does not contribute much towards controlling the flyrock. This is probably due the fact that the gas pressure is reduced due to "top air-decking" that in turn reduces the heaving effect.
vii. There is a general tendency of the upward flyrock to travel towards the back side of the free face. However, flyrock did not travel to a greater distance in all the simulated cases. Hence, a reasonable clearance in the backward direction of the free face is sufficient to avoid risk arising from flyrock.

Huandong et al. (2002) carried out a dynamic analysis of charge initiation and explosive expansion for describing the rock movement. They advocated that two-phase gas–solid theory should be adopted for explaining movement in blasts. They concluded that the longer the duration of gas phase, the more the acceleration and hence the more the flyrock distance. They considered the pressure equivalent to 0.125 ρ_0 d^2 and spherical particles, ignored the collision of particles and calculated the particle acceleration defined by Equation 4.35, using the gas response time as given in Equation 4.36, and deployed simple ballistic equation (Equation 4.8) to calculate flyrock distance.

$$\frac{du_p}{d_t} = -\frac{1}{\rho_r}\frac{d_p}{d_x} + \frac{1}{2}\frac{\rho_a}{\rho_r}\frac{d}{dt}\left(u_a - u_p\right) + \frac{u_a - u_p}{\Delta_t} \tag{4.35}$$

$$\Delta_t = \frac{\rho_r k_x^2}{18\eta_g} \tag{4.36}$$

where Δ_t is the gas response time, k_x is the particle or flyrock diameter, ρ_r is the particle density, ρ_a is the air density, u_a is the gas velocity, u_p is the particle velocity, η_a is the gas viscosity coefficient under movement, and p is the pressure.

The method deployed, though unique, has several issues in estimating the individual variables and gas properties that at the most can be approximated only.

An important study by Stojadinović et al. (2011) provided two approaches, viz. approximate numerical solution and application of Runge–Kutta algorithm to predict the flyrock distance. Their derivation was based on Newton's Second Law of Motion to determine the coordinates of a flyrock trajectory in space while considering three major vectors of flyrock travel, i.e. gravity (**G**), resistance force or drag (**D**), and force of lift (**L**). They defined the relationship $ma = \boldsymbol{G} + \boldsymbol{D} + \boldsymbol{L}$ as an equilibrium condition and derived a new concept for air drag (Equation 4.37).

$$b_d = \frac{2mg}{\rho_{air} A v_T^2} \tag{4.37}$$

where m is the mass of the flyrock fragment in kg, g is the gravitational constant in m/s^2, ρ_{air} is the density of air in kg/m^3, A is the cross-sectional area of the flyrock fragment in m^2 along a plane perpendicular to the direction of flight, b_d is the drag coefficient, and v_T is the flyrock terminal velocity in m/s.

4.6 INTELLIGENT METHODS

Artificial intelligence or neural networks (ANN) have been widely used for prediction of flyrock distance by numerous authors. The methods vary over a wide domain and incorporate different algorithms for prediction of flyrock distance, as compiled in Table 4.4. However, an overview of such publications is presented in Table 4.4.

The shortcomings of the data presented in Table 4.4 have been stated earlier. Although there is mention in some of the publications of selection of variables for prediction of flyrock using ANN method in Table 4.4, the rationale behind these is not explained. For example, out of 48 publications, 25 in Table 4.4 have maximum charge per delay (Q_{max}) in their prediction models, but the reason for inclusion of Q_{max}, a variable integral to prediction of ground vibrations due to blasting, is not explained. How Q_{max} is important as a variable in flyrock distance predictions is unclear and lacking logic. Unless charge per hole is not clearly stated to be equal to the Q_{max}, the rationale behind its use in flyrock distance prediction cannot be justified. Most of the models presented in Table 4.4 do not include the flyrock causative reasons of rockmass anomalies and inconsistencies and improper blast practice and others to explain the variability. The high degree of correlations in the models in terms of R^2 and least RMSE leave little room to explain the uncertainties. The classification of a particular fragment into flyrock is lacking in such studies as fragments landing at even 10 m have been considered as flyrock. There is no mention of air drag and other impact in the prediction in majority of the ANN models.

4.7 PRESSURE-TIME-BASED METHODS

The initial or exit velocity calculations is integral to flyrock distance prediction. There are few models that use the time interval of impact for prediction. However, the impact in terms of pressure exerted by high-pressure gases from blastholes has not been determined but estimated from approximate solutions using estimated blasthole pressure.

Raina et al. (2015a) presented a study that iterated the pressure measured at a known distance from blastholes in varied rock types. Based on such pressure calculations and through iterative means, the flyrock distance was calculated using a semi-empirical method. The prediction through this method was further corrected for air drag that resulted in a high degree of correlation between the observed and predicted flyrock. The methodology adopted is further explained in Figure 4.7.

A few controlled full-scale blasts were conducted by monitoring the dynamic, quasi-static, and static pressure in a dummy hole placed about one burden distance from blasthole in some selected mines with varied geomining conditions. The experiments were aimed at understanding the rock–explosive interaction by obtaining the pressure–time history at different stages of rockmass heaving and fragmentation. The pressure at the escape, corrected for probe distance, and corresponding time were used for iterations to arrive at flyrock distance to pick up the best pressure–time

TABLE 4.4
Details of the Artificial Intelligence Techniques Used for Flyrock Distance Prediction through Literature (in Chronological Order)

S. No.	Drill Diameter (mm)	Technique used	Case Count	Inputs—Rock, Drill, and Blast Design Variables	Output(s)	R_f (m)	Results R^2	Citations
1.	75–115	ANN; 8-3-3-2 topology	25	$d, l_{bh}, B/S, l_s, N_r, q, \rho_r, Q_{max}$	R_f, k_{50}	25–186	0.988	Monjezi et al. (2010)
2.	203	Fuzzy	490	$B, S, l_{bh}, b_{sd}, l_s, Q_{max}, \rho r, q$	R_f	10–70	0.985	Rezaei et al. (2011)
3.	76, 89, 115	ANN, FBPNN, multiple architectures tested	213	$B/S, Q_{max}, d, l_{bh}, l_s, b_{sd}, q, SMR, I_{BI}$	R_f	23–196	0.970	Monjezi et al. (2011)
4.	114.5–152.4	SVM	245	$d, l_{bh}, S, B, l_s, q, b_{sd}$	R_f,	25–95	0.970	Amini et al. (2012)
5.	42, 64, 76, 100	TOPSIS	19	d, l_{bh}, B, S, l_s, DP			–	Monjezi et al. (2012b)
6.	75–115	NG-ANN	195	$d, l_{bh}, S, B, l_s, q, b_{sd}, Q_{max}, RMR$	R_f, l_o	*	0.976	Monjezi et al. (2012a)
7.	–	ML-SVM	234	$l_{bh}, S, B, l_s, q, b_{sd}$	R_f	10–100	0.948	Khandelwal and Monjezi (2013)
8.	115, 165	ANN	30	q_{lc}, B, l_s, q, UCS, RQD	R_f	20–56	0.983	Trivedi et al. (2014)
9.	89–150	PSO based ANN	44	$d, l_{bh}, Q_{max}, S, B, l_s, q, \rho_r, l_{sub}, N_r$	R_f, v_{max}	71–410	0.930	Armaghani et al. (2014)
10.	115	ICA-ANN	–	$l_{bh}, B/S, l_s, Q_{max}, q, \rho_r$	R_f	44–206	0.980	Marto et al. (2014)
11.	127	ANN-Fuzzy	230	$l_{bh}, B, S, l_s, q, Q_{max}$	R_f	20–100	0.939 (ANN) 0.957 (Fuzzy)	Ghasemi et al. (2014)
12.	115, 165	ANFIS	100	$Q_h, q_{lc}, l_{bh}, B, S, l_s, q, d, \sigma_c$, RQD	R_f	54–222	0.98	Trivedi et al. (2015)
13.	75, 89, 115, 150	ANN, ANFIS	232	Q_{max}, q	R_f, v_{max}, p_{oa}	72–328	0.925 (ANN) 0.979 (ANFIS)	Armaghani et al. (2016)

S. No.	Drill Diameter (mm)	Technique used	Case Count	Inputs—Rock, Drill, and Blast Design Variables	Output(s)	R_f (m)	Results R^2	Citations
14.	–	GP, GEP	97	B, S, l_s, l_{bh}, q	R_f	67–354	0.935 GP 0.991 GEP	Faradonbeh et al. (2016)
15.	–	ACO	97	B, S, l_{bh}, l_s, q	R_f, l_{oa}	200–300	0.994	Saghatforoush et al. (2016)
16.	80, 105, 130, 155	GP	262	$q, l_s, B/S, d, Q_{max}, l_{bh}$	R_f	60–405	0.908 GP 0.816 NMLR	Faradonbeh et al. (2016)
17.	76/110/250	ANN	36 (908 holes)	$d, l_{bh}, \alpha_{bi}, l_s, B, S, t_{RR}, N_{ff}, Q_{max}, \rho_e, c_d, V_g$	v_0, R_f	*		Stojadinović et al. (2016)
18.	100, 152	BPNN	334	$d, l_{bh}, H_b, l_{sub}, N_h, S, B, Q, l_s, q, b_{sd}, t_{HH}$	R_f	10–100	0.97	Yari et al. (2016)
19.	115, 165	BPNN	20	$Q, q_l, l_{bh}, B, S, l_s, q, d, \sigma_c$, RQD	R_f	50–250	0.98	Trivedi et al. (2016)
20.	75, 115, 150	PSO	76	S, B, l_s, q *and* ρ_r	R_f	61–334	0.966 MLR 0.988 PSO	Hasanipanah et al. (2017b)
21.	–	DT/CART; MLR	65	$l_{bh}, S, B, l_s, q, Q_{max}$	R_f	72–328	0.77, MLR 0.946 DT	Hasanipanah et al. (2017a)
22.	76, 152	DE+DA	30	$d, l_{bh}, N_h, S, B, Q, l_s, q, b_{sd}$	R_f	25–100	Err DE-15, DA-7%	Dehghani and Shafaghi (2017)
23.	–	DA+FIS	320	$B, S, N_h, Q_{max}, t, RMR, l_{bh}, q, l_s$	R_f	10–100	0.976	Bakhtavar et al. (2017
24.	75–150	GEP	76	d, S, B, l_s, q	R_f	60–405	0.92	Faradonbeh et al. (2018)
25.	251	LS-SVM	90	$B/S, H_t/B, l_{sub}, l_s, Q_{max}, \rho_r, q$	R_f	10–70	SVM-0.945 LS-SVM-0.969	Rad et al. (2018)
26.	76	MHA FFA	200	$B, S, l_{bh}, l_s, q, Q_{max}$, GSI	R_f	75–280	0.94	Asl et al. (2018)

(Continued)

TABLE 4.4 (Continued)
Details of the Artificial Intelligence Techniques Used for Flyrock Distance Prediction through Literature (in Chronological Order)

S. No.	Drill Diameter (mm)	Technique used	Case Count	Inputs—Rock, Drill, and Blast Design Variables	Output(s)	R_f (*m*)	Results R^2	Citations
27.	75, 89, 115, 150	ICA-ANN, GA, PSO	262	B/S, d, q, l_s, Q_{max}, l_{bh}	R_f	60–405	ICA 0.943, PSO 0.958 GA 0.930	Koopialipoor et al. (2019)
28.	–	ICA	78	B, S, l_s, Q, RMR	R_f	96–267	ICA-L 0.954	Wu et al. (2019)
29.	75–150	MRA-ANN, MC	260	d, l_{bh}, B/S, l_s, Q_{max}, q	R_f	67–354	MRA 0.819 ANN 0.954	Zhou et al. (2020a)
30.	–	FRES; Risk Optimization GA, ICA, PSO	62	B, Q_{max}, q, S/B, l_s/B, Hb, α_{bi}, c_d, d, B/d, RMR	R_f	100–325 *inf.*	GA 0.984 PSO 0.981 ICA 0.983	Hasanipanah and Amnieh (2020)
31.	75,89,115,150	PCR, SVR, MARS Opt. GWO	262 52 data given	B/S, d, q, l_s, Q_{max}, l_{bh}	R_f	150–240 *inf.*	PCR 0.642 MARS 0.884 MARS 0.939	Armaghani et al. (2020)
32.	75,115,150	ELM, OR-ELM	82	S, B, l_s, q, ρ_r	R_f	61–334	LReg 0.883 ANN 0.912 ELM 0.955 OR-LEM 0.958	Lu et al. (2020)
33.	75–150	ELM, BBO	262	B/S, d, q, l_s, Q_{max}, l_{bh}	R_f	67–354	PSO-ELM 0.93 BBO-ELM 0.94 ELM 0.79	Bhatawdekar et al. (2020)
34.	Ullu-Tirram	ANN, PSO	65	l_{bh}, Q_{max}, B, S, l_s, q	R_f	71–328	ANN 0.9	Zhou et al. (2020b)
35.	97.5, 127.5	RF, BN	262	d, l_{bh}, B/S, l_s, q	R_f	67–354	-	Han et al. (2020)

S. No.	Drill Diameter (mm)	Technique used	Case Count	Inputs—Rock, Drill, and Blast Design Variables	Output(s)	R_f (m)	Results R^2	Citations
36.	–	RFNN+GA	70	S, B, l_s, Q_{max}	R_f	100–280 *inf.*	ANN 0.866 GA-ANN 0.944 NLMR 0.877	Rad et al. (2020)
37.	–	RFNN—PSO	72	Q_{max}, l_s, B, S	R_f	94–269	ANFIS 0.876 RFNN-PSO 0.974	Kalaivaani et al. (2020)
38.	–	DNN, WOA (opt.)	240	$l_{bh}, Q_{max}, B, S, l_s, q$	R_f	71–328	0.9829	Guo et al. (2021)
39.	75, 115, 150	ANFIS-CA ANFIS-GOA	80	q, l_s, ρ_r, S, B	R_f	61–334	ANFIS-GOA 0.974 ANFIS-CA0.953	Fattahi and Hasanipanah (2021)
40.	76, 152	GEP, CO	268	B, S, l_s, lq, q	R_f	30–100	GEP 0.91	Dehghani et al. (2021)
41.	76, 102, 150	MLP, RF, SVM, HHO-MLP, WOA	152	d, GSI, I_w, RQD, l_{bh}, $l_s/B, q_{lc}, q$	R_f	103–549	MLP 0.953 RF 0.933 SVM 0.937 HHO-MLP 0.991 WOA 0.972	Bhatawdekar et al. (2021)
42.	75,115,150	KELM, FRD, LWLR, RSM, BRT	73	S, B, l_s, q	R_f	109–334	KELM 0.970 LWLR 0.976	Jamei et al. (2021)
43.	76, 110, 250	N-N	4045 holes 908 flyrock	$d, l_{bh}, \alpha bi, ls, ls/Q, l_s/l_{bh}$, $B, S, V_h, q, t, N_{ff}, Q$, ET, ρ_e, c_d, $V_g, \rho_r, \sigma_c, st$	R_f	50–580	N	Stojadinović et al. (2021)
44.	–	LMR, GEP, optimization	152	B, S, l_s, q	R_f	85–190	LMR 0.86 GEP 0.91	Monjezi et al. (2021)
45.	75, 115, 150	ANN-ADHS, HS, PSO	82	S, B, l_s, q, ρ_r,	R_f	61–334	ANN-ADHS 0.929 HS 0.871 PSO 0.832	Hasanipanah et al. (2022)

(Continued)

TABLE 4.4 (Continued)
Details of the Artificial Intelligence Techniques Used for Flyrock Distance Prediction through Literature (in Chronological Order)

S. No.	Drill Diameter (mm)	Technique used	Case Count	Inputs—Rock, Drill, and Blast Design Variables	Output(s)	R_f (m)	Results R^2	Citations
46.	76, 140	ICA, ANFIS, ANN	282	d, B, l_s, q, Q_{max}	R_f	13–100	0.956	Shakeri et al. (2022)
47.	76, 114, 127	FCM; ACWNNsR,	416	B, S, l_s, q, Q_{max}	R_f	46–389	0.996	Hosseini et al. (2022)
48.	75–150	EMLM/BN	262	$l_{bh}, B/S, l_s, Q_{max}, q$	R_f	67–354	0.974	Barkhordari et al. (2022)

Note: H_b is the bench height, B is the burden(m), S is the spacing (m), l_{bh} is the hole depth (m), b_{sd} is the specific drilling (m/m^3), l_o is the overbreak (m), l_s is the stemming (m), Q_{max} is the charge per delay, q is the specific charge, ρ_r is the rock density ($\times 10^{-3}$ kg/m^3), ρ_e is the explosive density ($\times 10^{-3}$ kg/m^3), d is the blasthole diameter, V_h is the volume of blasthole, V_g is volume of gases, α_{bi} is the blasthole inclination, N_{ff} is the number of free faces, c_d is the velocity of detonation of explosive, t is the hole-to-hole delay (time), σ_c is the compressive strength of rock, N_h is the number of holes in a blast, N_r is the number of rows in a blast round, s_t is the stemming factor, I_w is the index of weatherability, *RMR* is the rockmass rating, GSI is the geological strength index, *SMR* (not explained by the authors), I_{BI} is the index of blastability, and *B/S* is the ratio of burden to spacing.

Acronyms: DP: drilling pattern (rectangular or staggered), ET: explosive type, FBPNN: feedforward back propagation neural network, Fuzzy: fuzzy logic, SVM: support vector machine, LOO: leave one out cross-validation method, TOPSIS: technique for order preference by similarity to ideal solution, NG-ANN: neurogenetic artificial neural network, ML: machine learning, PSO: particle swarm optimization, ICA-ANN: imperialist competitive algorithm, ANFIS: adaptive neuro fuzzy inference system, GP: genetic programming and GEP: genetic expression programming, ACO: ant colony algorithm, DT: decision tree, CART: classification and regression tree, DE: differential evaluation, EMLM/BN: ensemble machine learning method/Bayesian network, algorithm, DA: dimensional analysis algorithm, LS-SVM: least squares support vector machines, GSI: geological strength index, MHA: metaheuristic algorithm, FFA: firefly algorithm, SEM: probabilistic structural equation model, FRES: fuzzy rock engineering system, PCR: principal component regression, SVR: support vector regression, MARS: multivariate adaptive regression splines, GWO: grey wolf optimization, ELM: extreme learning machine, ELM: outlier robust (OR-ELM), ELM: extreme learning machine, BBO: biogeography-based optimization, RF: random forest, BN: Bayesian network, RFNN: recurrent fuzzy neural network, GA: genetic algorithm, DNN: deep neural network, WOA: whale optimization algorithm, GOA: grasshopper optimization algorithm, CA: cultural algorithm, GP/GEP: gene expression programming, COA: cuckoo optimization algorithm, KELM: kernel extreme learning machine, LWLR: local weighted linear regression, RSM: response surface methodology, BRT: boosted regression tree, LMR: linear multiple regression, ANN-ADHS: ANN coupled with adaptive dynamical harmony search, ANN-HS: hybrid ANN models coupled by harmony search, FCM: fuzzy cognitive map.

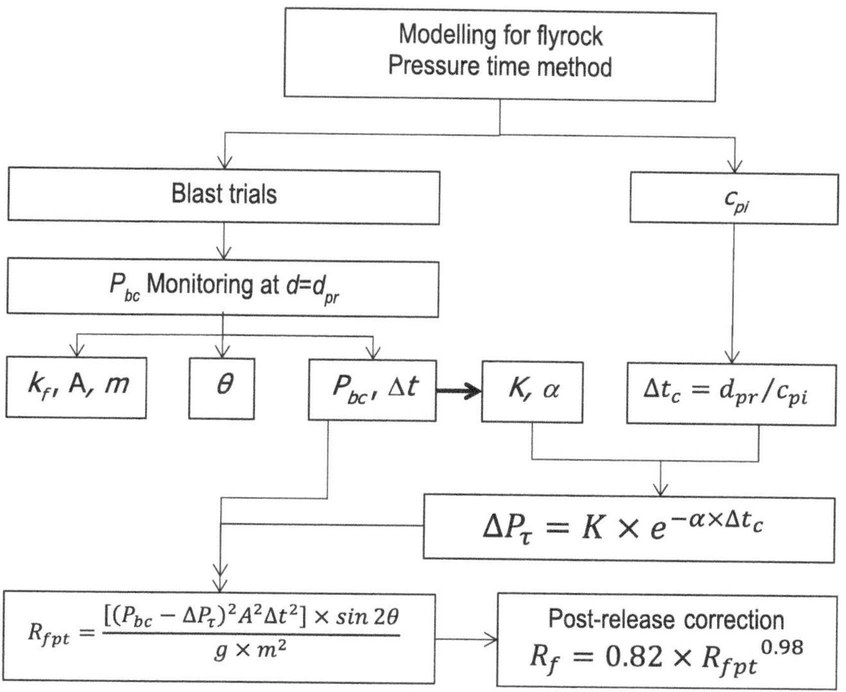

FIGURE 4.7 Methodology for calculation of flyrock distance using pressure–time technique.

combination matching the observed flyrock. The picked-up pressure–time values were used in the equation of trajectory motion to estimate the flyrock distance along with other known values of the flyrock fragment. The results of the flyrock prediction were tested with known cases with known escape pressure.

The equation for flyrock prediction from pressure–time method is given in Equation 4.38:

$$R_{fpt} = \frac{\left[\left(p_{bc} \pm \Delta p_\tau\right)^2 A^2 \Delta t^2\right] \times \sin 2\theta}{g \times m^2} \tag{4.38}$$

where R_{fpt} is the distance of flyrock in m predicted by using pressure–time method, p_{bc} is the pressure acting on the flyrock fragment at escape in Pa, and Δp_τ is the adjustment for pressure in Pa for probe distance with respect to burden B.

Two conditions for extrapolation of p_{bc} are as follows:

1. Pressure probe distance >B, then Corrected $p_{bc} = p_{bc} + \Delta p_\tau$
2. Pressure probe distance <B, then Corrected $p_{bc} = p_{bc} - \Delta p_\tau$

where Δp_τ can be deduced from p_{bc}–t correlation, A is the area of fragment in m^2, Δt is the time of application of the pressure on the flyrock fragment in s, g is the acceleration due to gravity in m/s^2, and m is the mass of the flyrock fragment in kg.

The results of the investigations in nine trials with controlled change of rock and explosive variables provided better results for flyrock prediction in comparison to the empirical model. Thus, the pressure–time method with suitable correction can be used for direct estimation of flyrock distance after application of correction for post-release (from bench) effects, including air drag, Magnus effect, and rebound. The post-release correction worked out from model blasts was applied to Equation 4.38 to yield the final flyrock distance prediction model as given in Equation 4.39:

$$R_f = 0.82 \times R_{fpt}^{0.98} \tag{4.39}$$

Equation 4.38 can be calibrated for parameters in particular mining condition through experimental trials.

4.8 SOME INSIGHTS INTO FLYROCK PREDICTION

4.8.1 Concrete Models

Concrete models tests, as introduced in Section 2.3, were conducted by the author to investigate the flyrock occurrence and to identify the conditions responsible for generation of flyrock. The tests were conducted on concrete models of different compressive strengths, artificial jointing, drill diameters, explosive, and burden variations. The objective of tests, despite the associated shortcomings, was to have insights into the following issues:

1. What should be called a flyrock? A categorization of flyrock distance in terms of the drill depth considering the same to be equal to bench height. This should answer the probabilities of flyrock if low distance fragments are eliminated from the analysis. Also, a foundation could be laid for actual field tests for considering a fragment as flyrock or not depending upon its distance from blast and bench height.
2. Which fragments moving independently should be classified as flyrock? Should it be independent of bench configurations like height or width, or not?
3. Quantification of error from predictions by ballistic trajectory method and delineating the underlying reasons for such errors. Some specific inferences have already been discussed in Section 4.2.
4. In normal conditions, how far can a flyrock travel and what variables should be considered in flyrock distance prediction?
5. What is the influence of design and joint variables on the flyrock distance?
6. Quantification of errors due to rebound of fragments after falling on ground.

The ratio of flyrock distance (R_f) to drill depth l_{bh} (Figure 4.8) shows that the rock fragments in such test blasts travelled up to 28 drill depths. This analysis presents a method to discriminate between throw, excess throw, and flyrock and will be extended

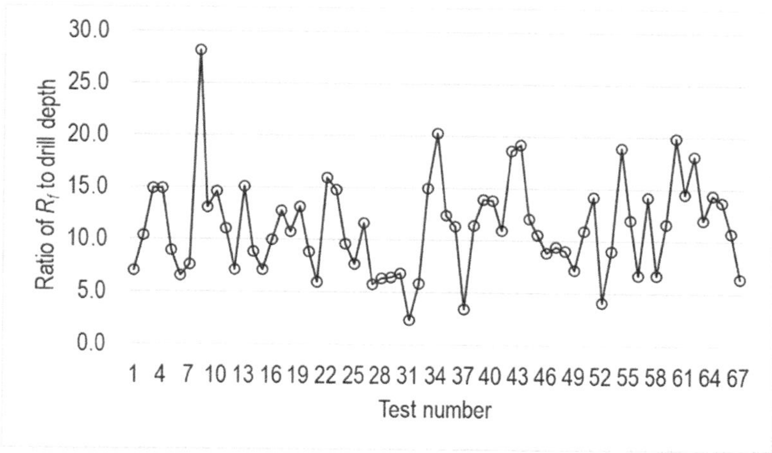

FIGURE 4.8 Ratios of flyrock distance to the drill depth in model blasts.

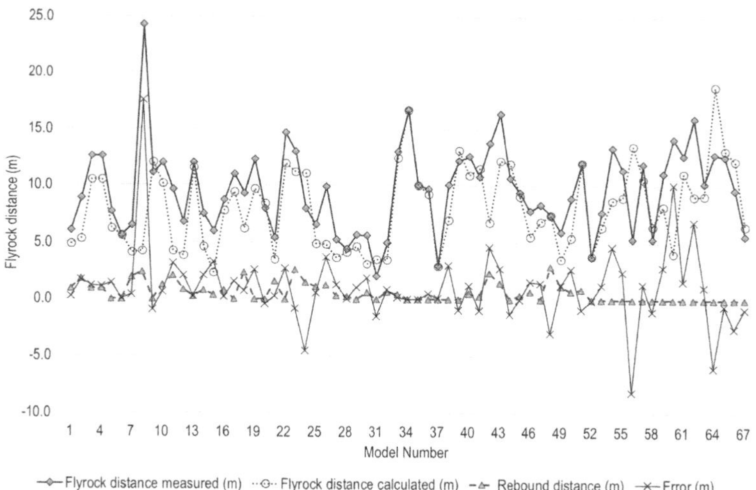

FIGURE 4.9 Comparison of flyrock distances after adding rebound distances; error mentioned in the figure is the difference between the flyrock distance measured and the distance after adding rebound distance to the initial calculations of flyrock using trajectory equations. Negative values indicate flyrock movement towards the backside of the block.

to actual field data in Section 4.8.3. It is apparent from the plot that a ratio of R_f to l_{bh} of around 5–7 can be in the category of throw or excess throw. Beyond this ratio, all fragments can be treated as flyrock. However, the ratio needs to be transformed or controlled with bench width that presents a better rationale for flyrock projection.

A comparison of predicted and observed fragment travel distance from the test blasts (Figure 4.9) shows that there is significant error in predictions using ballistic trajectory method while calculating the initial velocity of such fragments from high-speed video and launch angles.

A summary of such errors and reasons thereof compiled in Table 4.5 shows that there is rebound in significant number of cases, i.e. in 49% cases the fragments on touching the ground rebounded. The other reasons for error in prediction is air drag and dragging of fragments on the ground even after rebound and fall.

The errors in prediction are quite significant and hence the ballistic equation does not fit into the scheme of things properly (Figure 4.10) unless other factors mentioned in Table 4.5 are accounted for. One more observation from the model tests is that the launch angle of flyrock does not necessarily relate to its maximum travel distance (Figure 4.11) and the trajectory is controlled by several other design and

TABLE 4.5
Details of Tests on Concrete Blocks along with Errors and Reasons for Error

Test and Error Inferences		Measure	Statistics
Method used	Ballistic trajectory method	No. of tests	67
Calculation error	Camera and computing	Rebound events recorded	49%
Other reasons for prediction error	• Air drag • Rebound (identified from videos) • Dragging of fragment along the surface after falling but no rebound	Average error in initial calculated travel distance Error after inclusion of rebound distance	28% 20%

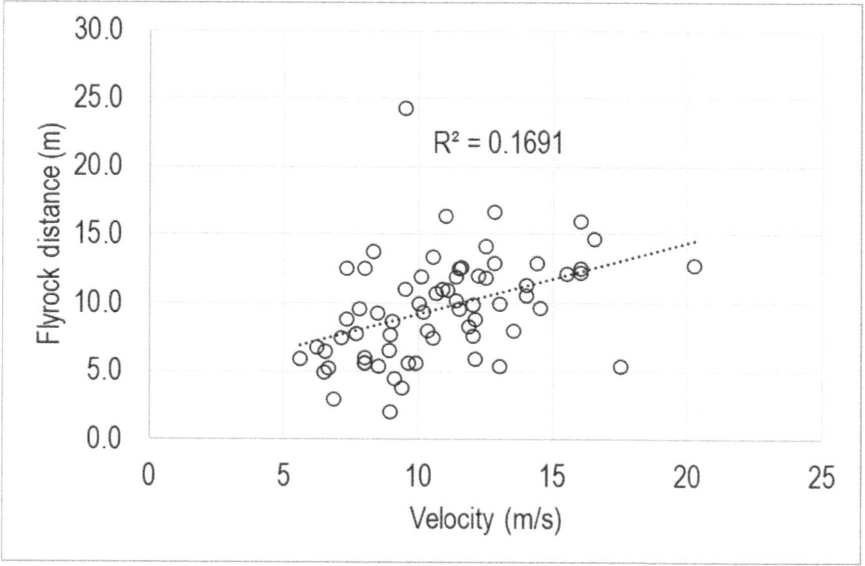

FIGURE 4.10 Flyrock distance versus flyrock velocity as observed in the concrete model test blasts.

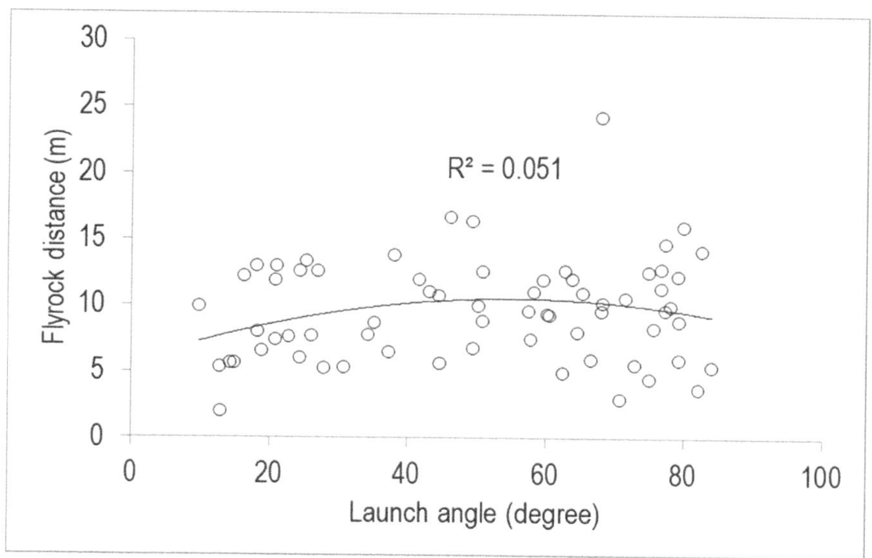

FIGURE 4.11 Flyrock distance versus launch angle of flyrock as measured in the concrete model test blasts.

rock variables. However, a feeble trend of maximum flyrock distance at around 45° was observed in the plot (Figure 4.11).

The model tests showed that there is some influence of blast design variables on the flyrock distance. Around 35% of the fragments were thrown towards back side of the models and just 10% on the sides of the models. The results of the experiments in terms of burden, an important design variable, and joints in the models are summarized in Table 4.6 that shows the correlations of flyrock distance (R_f) with individual variables by one factor at a time method while eliminating points that were thrown less than 7 m from the blocks, and considering the rest of the values as flyrock.

One can find that there is influence of the type of material being blasted, burden, and stemming with the flyrock distance. However, this does not completely justify the occurrence of flyrock. It may be mentioned that the models were designed to produce flyrock and the probabilities of flyrock occurrence are high in this case which may not represent the actual field conditions. Despite that, there were several models where the fragments thrown cannot be classified as flyrock.

Sphericity of the flyrock(s) from model tests, calculated by the method of Sneed and Folk (Sneed & Folk, 1958) and plotted with the help of Triplot© (Graham & Midgley, 2000),[1] is provided in Figure 4.12. The sphericity of the flyrock(s) shows maximum occurrence beyond sphericity value of 0.49 with a maximum in the range of 0.6–0.71 (Figure 4.12) that corresponds to the compact–elongate shape of the particles. The finding can be useful in future research for estimating the air drag and post-ejection movement of the flyrock with proper calibration with data from field blasts. However, the distribution of flyrock in

TABLE 4.6
Correlation of Different Variables and Their Strength With Respect To Flyrock Distance

Variable	R^2 with R_f (Trend)	Variable	R^2 with R_f (Trend)	Variable	R^2 with R_f (Trend)
Rock factor	0.32 (+)	Burden	0.57 (−)	Stemming	0.11 (−)
Specific charge	0.24 (+)	Explosive length to blasthole volume	0.56 (+)	Rock factor has been defined in terms of strength of the block, density of the block, block size, and the nature of joints	

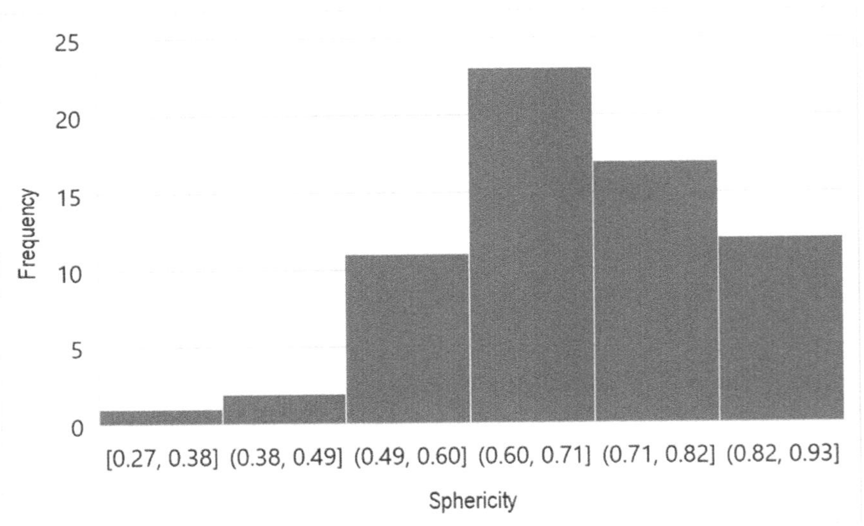

FIGURE 4.12 Histogram of the sphericity of the flyrock from the measured dimensions.

sphericity represents a wide range from 0.27 to 0.93 that points to the fact that flyrock can assume any shape.

4.8.2 Field Testing of Flyrock

Three hundred and sixty-three field data from the actual blasts (Table 4.7), generated by the author, have been investigated for determination of feasible methodology for prediction of flyrock distance. The data includes a wide range of rock types of sedimentary, metamorphic and igneous origin and covers a wide range of rockmass– and flyrock-related conditions, drill diameters, and blast design variables. This includes the typical case of flyrock discussed in Section 3.5.1 where blasting had been stopped by the regulatory authorities.

TABLE 4.7
Brief Details and Statistics of the Data for Flyrock Evaluation (All Units Are in SI)

Rock Type	No. of Tests	Other Details	Stats Min	Stats Max	Stats Avg	Other Details	Stats Min	Stats Max	Stats Avg
Iron ore/laterite	2	Drill diameter	0.1	0.165	0.123	Density of rock	1800	3,100	2400
Granite and Cu ore	4	Bench height	0.9	14.0	6.8	Density of explosive	700	1,100	1000
Mn ore	12	Blasthole depth	0.9	14.5	7.1	Charge per hole	0.3	180.3	36.9
Granite OB	2	Burden	0.7	5.5	2.9	Charge per blast	8	11,022	951
OB limestone/reject	8	Spacing	1.2	9.4	3.8				
OB Mn	4	Stemming length	0.4	6.0	3.2	Specific charge	0.1	2.1	0.4
Schist, Gneiss, +Mn ore	11	Charge length	0.1	10.0	3.6	Throw	1	28	12
Cu ore, quartzite	15	Average joint spacing	0.1	1.2	0.4	Flyrock >40 m	40	250	72
Limestone	296	Flyrock >30 m	30	250	58	S to B ratio	0.8	3.7	1.3
Waste granite	9	Flyrock size	0.03	0.61	0.19	l_s to B ratio	0.3	5.1	1.2
Total No. of blasts	363	l_q to l_{bh} ratio	0.1	0.8	0.5	l_{bh} to B ratio	0.6	7.8	2.4

As discussed in Section 4.8.1, a need exists to differentiate between throw, excess throw, and flyrock so that a plausible analysis of the flyrock distance could be attempted. A simple solution, based on the inferences from concrete model blasts, could be achieved for the field blasts by defining the limits of such displacements of the blasted material and fragments projected beyond the throw and excess throw conditions (Figure 4.13). The following scheme is developed and proposed for adoption while recording flyrock, along with its other attributes:

1. Throw (R_T) does not exceed the bench width.
2. Excess throw is limited by two bench widths.
3. Flyrock: rock fragments falling beyond the excess throw.
4. Probabilities of flyrock occurrence can be then defined in a logical manner based on the aforementioned criteria.

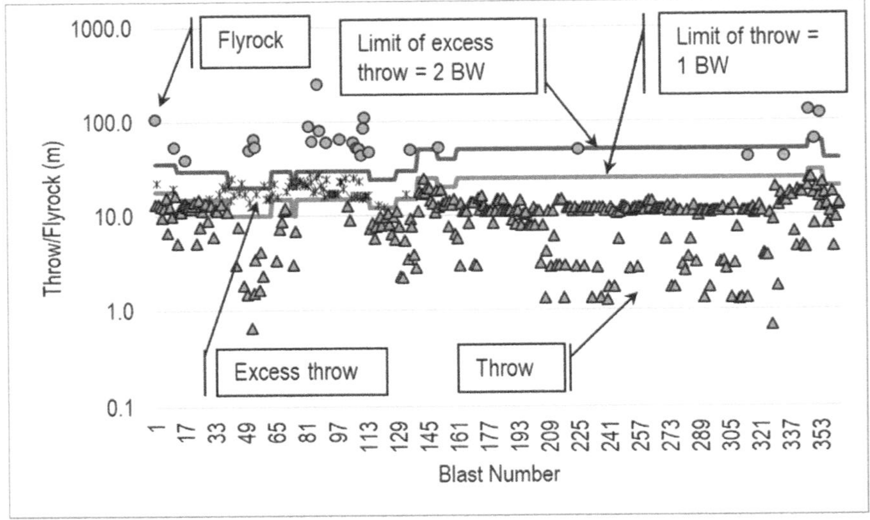

FIGURE 4.13 Differentiation method between throw, excess throw, and flyrock distances.

This means that any fragment falling within one bench width should not be classified as flyrock. A simple reason for this is that if the bench topples on its own without blasting, the broken material will fall up to a distance that is more than the bench height. A clear demarcation can be made to identify the flyrock in this process as demonstrated earlier. The methodology suggested in Figure 4.13 is expected to provide a basis for refining the definition of flyrock distance in terms of bench width, help to develop universal prediction models, and at the same time help in defining the probabilities of flyrock occurrence.

4.8.2.1 Modelling Constraints

- It is practically not possible to incorporate rock anomalies in modelling. This demands the use of correction factor.
- The major influencing design variable on flyrock is the stemming length, type, etc. If flyrock occurs from the stemming zone, the preposition amounts to poor blast design and poor engineering practice.
- Due to excessive spacing and burden, the rock may break in the case of highly jointed formations, but the movement of material and hence flyrock is towards the top of the face. If flyrock occurs from this side, the accounting and calculations take a different turn with respect to spacing and burden.
- The throw of the fragmented rockmass is by design and a regular phenomenon but flyrock is by chance and not regular in nature.
- Measurement of the velocity and launch angle of the flyrock are practically of relative nature owing to 2D nature of the videos. This influences the measurement of true direction of flyrock emanating from the face. Taking

queue from the physical model experiments, and the controlled nature of such tests, one can have a fair idea of the probabilities of the flyrock in different directions.

Considering the classification of throw, excess throw, and flyrock occurrence (Figure 4.13), it is observed that the probabilities of flyrock change. This distinction is essential to put all the recent publications into perspective, particularly relating to prediction of flyrock distance using intelligent techniques, wherein all distances of fragment displacements due to blasting have been incorporated as flyrocks. It may not be out of place to mention that flyrock is an aberration and clubbing it with throw is contradictory to basic premise of prediction. Hence, just using blast design variables as components defining flyrock may not be logical as statistically it means clubbing regular data with outliers. A simple construct of the earlier prediction mechanism (Raina et al., 2006a) has been tried over here to explain the logic behind flyrock prediction.

4.8.3 Modelling the Unknown: Flyrock

As explained earlier, fragment size, throw, ground vibration, and air overpressure associated with blasting can be modelled with ease as these are regular and essential variables in blasts. Each blast produces a definite fragment size over a defined distribution, throw of the broken rockmass that may be negligible to order of few bench heights, ground vibrations that are generated depending on maximum charge per delay, and attenuate over distance, and same is the case with air overpressure. This provides a firm basis for their prediction. However, in the case of flyrock, its distance is different than the throw, but both are guided by the same mechanism and blast design variables. This requires building a firm basis for predictive mechanism for flyrock while taking the following premise into account:

1. A blasthole has a zone of influence determined by its burden, spacing, and drill length with breakage limited to the bench height. Hence, there is a volume of rock that is to be fragmented and thrown. The product of the density of the rockmass and the volume being blasted, thus represents the load.
2. The blasthole is loaded with a specific quantity of explosive (M_e) and has a density that determines its impedance and represents the energy available for the work.
3. The ratio of the weight of the rock being blasting (M_r) and the weight of explosive (M_e) represents the balance of forces.
4. Since both the density of the rockmass (ρ_r) and the density of the explosive (ρ_e) are considered, it represents a dynamic ratio, as density of rock has a relationship with the p-wave velocity and density of the explosive relates to its velocity of detonation.
5. The stiffness of the bench is determined by the ratio of bench height (H_b) and burden (B) that has a significant role in throw and fragmentation during blasting (Konya, 1995).

6. Degree of jointing influences the level of work that explosive in a blasthole must perform to break and throw the rockmass. Average joint spacing (S_j) is a measure of indirect confinement of explosive gases and represents the in situ block size and properties of the rockmass. It is important to note that the higher the joint frequency, the higher the chances of faster escape of explosive gases, and vice versa. It will be prudent to recall numerical analysis results in Section 4.5 that are coherent with the aforementioned statement.

A combination of the factors mentioned earlier can be used to define the throw of a blast (R_T) as follows (Equation 4.40):

$$R_T = f\left(S_j, M_r, M_e, \frac{H_b}{B}\right) \text{ or a simple approximation will be,}$$
$$R_T = S_j \times \frac{M_e}{M_r} \times \frac{H_b}{B} \tag{4.40}$$

where S_j is the average joint spacing of rockmass, M_r is the mass of rock to be blasted in a hole, M_e is the mass of explosive in a hole, H_b/B is the stiffness ratio, i.e. ratio between bench height and burden.

The relationship (Equation 4.40) is a non-redundant one which has in-built combination of host of variables and presents a basis in terms of ratio of contributing and resisting forces in a blast. There are possibilities of further simplification of the concept presented in Equation 4.40. We can call the right-hand side of Equation 4.40 as "throw factor." The relationship (Equation 4.40) attempted on the data cited earlier and plotted against the throw of the rockmass during blasting is represented by Figure 4.14, which gives a fair idea of the significant correlation of throw of the blasted muck with the throw factor.

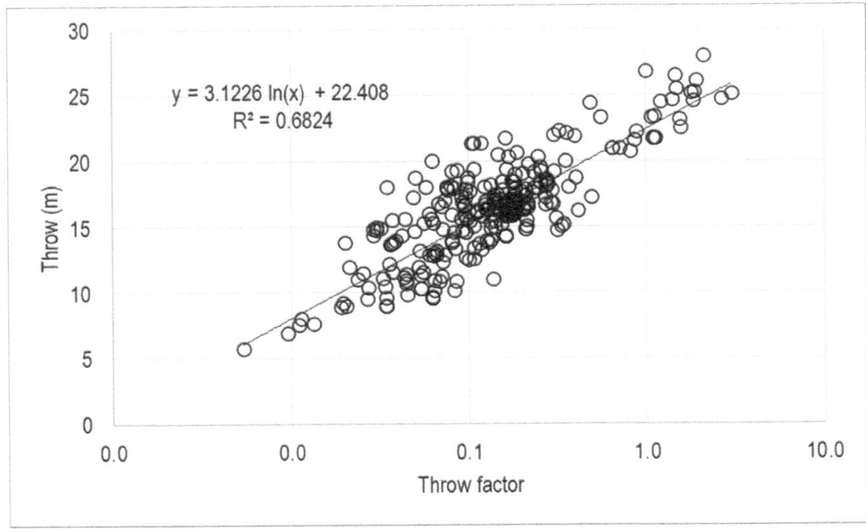

FIGURE 4.14 Relationship of throw of the broken rockmass with the throw factor.

Equation 4.40 shows a fair correlation with throw ($R^2 = 0.68$) and can be useful in predicting throw or even excess throw of the fragmented muck. Equation 4.40 was further resolved for interactions in component variables through multivariate analysis (ANOVA) by utilizing the data in Table 4.7. The ANOVA revealed the interactions in variables that yielded an $R^2 = 0.85$, adjusted $R^2 = 0.84$, and predicted $R^2 = 0.83$ with an adequate precision (signal-to-noise ratio must be >4) of 83.07. The agreement in adjusted and predicted R^2 values, i.e. a difference of <0.2, indicates that the model is significant. Further results of the analysis are provided in Table 4.8.

The p-values of all the factors and interactions (Table 4.8) are also quite significant (<0.05). The contribution of the independent variables and other factors (Table 4.8) provides an insight into the throw of the blasted material as a function of the weight

TABLE 4.8
Multivariate Regression Analysis of Throw

Variable ▶ Stats ▼ |	M_r	M_e	H_b/B	S_j	$M_r \times S_j$	$M_e \times H_b/B$	$H_b/B \times S_j$	M_e^2	$(H_b/B)^2$
p-Value	0.0057	<0.0001	<0.0001	0.0049	<0.0001	0.0473	<0.0001	0.0001	0.0137
Contribution	5	13	24	5	14	9	16	9	5

M_r is the mass of rock to be blasted in a hole $\times 10^3$ kg, M_e is the mass of explosive in a hole, H_b/B is the stiffness ratio, i.e. ratio between bench height and burden, S_j is the average joint spacing (can be replaced by in situ rock blocks size for better results).

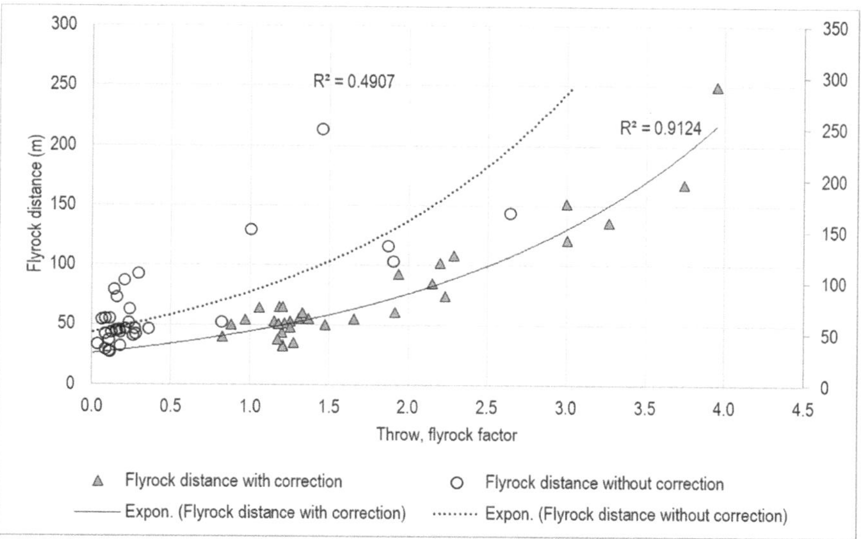

FIGURE 4.15 Correlation of flyrock distance with factor for flyrock with and without correction.

of rock and joint spacing, weight of rock, bench stiffness and joint spacing, and a higher order interaction of bench stiffness.

It is imperative from the analysis of throw that the prediction capability of throw can be quite accurate through such modelling. The model performance can be further fine-tuned by incorporating the in situ block size of the rockmass instead of average joint spacing. However, same is not the objective over here. The main concern is to identify whether Equation 4.40 can predict the flyrock distance.

The factors used in Equation 4.40 if treated in combination and designated as throw factor when treated with flyrock distance yields an insignificant correlation of $R^2 = 0.49$ (Figure 4.15, dotted trend line). The result indicates that mere blast design, rock properties, and explosive properties are not able to predict the flyrock distance in totality. The main reason for this is that flyrock is a result of anomalies and not integral to the blast design. Hence there is a need for a correction factor for different conditions or causative factors resulting in flyrock in such equation. For the cases under discussion, the correction factor (C_f) for flyrock is introduced as detailed in Table 4.9.

The values of the C_f when applied to the flyrock distance data observed improves the correlation to an R^2 of 0.91. The correction factor values (Table 4.9) may be specific to the mines and will require calibration to have a site-specific prediction model for flyrock distance.

TABLE 4.9
Inferred Values of the Correction Factor for Flyrock Distance Prediction from the Data Acquired by the Author

Causative factor	Correction Factor (C_f)	Controlling Factor	Correction Factor (C_f)
Top hole initiation	1.3	Bottom hole initiation using non-electric Shock tube combination	0.8
Stemming issues (length)	1.2	Electronic detonators (needs further validation)	0.7
Stemming type/method issues	1.2	Stemming	
Weak rock in stemming zone, previous blast damages	1.3	Rock chips	0.9
Excessive burden/choked blasts	1.3	Mechanical stemming or proper tamping	0.8
Rock conditions		Stemming plugs (needs further validation)	0.8
Weak rock in competent rock	1.3	Autonomous drilling	0.8
Weak joints/fractures in the face, previous blast damage indented face	1.2	Trimmed blast face, excellent face conditions	0.7
Slope blasting	2.0	Design match with drilling	0.8
Delay malfunction	2.5		
Boulder blasting/toe blasting	3.0		

The C_f (Table 4.9) called the flyrock correction factor, to be applied to Equation 4.40, not only provides for the causes of the flyrock but also accounts for the control measures. A general form of the flyrock distance prediction model is presented in Equation 4.41. The constants appearing in the Equation 4.41 will need to be determined for a particular mine.

$$\sqrt{R_f} = Int. + k_1 M_r - k_2 M_e - k_3 \frac{H_b}{B} - k_4 S_j - k_4 C_f - k_5 M_r M_e + k_6 M_r C_f + k_7 M_e \frac{H_b}{B} + k_7 S_j C_f + k_8 M_e^2 \tag{4.41}$$

where *Int.* is the intercept of the equation, k_1 to k_8 are coefficients of different quantities in the equation, M_r is the mass of rock to be blasted in a hole $\times 10^3$ kg, M_e is the mass of explosive in a hole in kg, H_b/B is the stiffness ratio, i.e. ratio between bench height and burden in m, and S_j is the average joint spacing of rockmass.

The method can be further improved to derive the relative weights of each variable in Equation 4.40 and calibration of the correction factor as given in Table 4.9. Numerical and pressure-based methods, as discussed earlier, can also be deployed for such calibration.

4.8.4 Flyrock Size

Flyrock size can be anything ranging from few centimetres to a few metres in diameter. There are evidences that flyrock size has been predicted, e.g. Lundborg (1974) and McKenzie (2009). Armaghani et al. (2014) reported flyrock fragment size of 0.035–0.058 m. Few authors have touched upon the subject of predictability of flyrock size, but the basis is very poor, as it is practically impossible to know the size

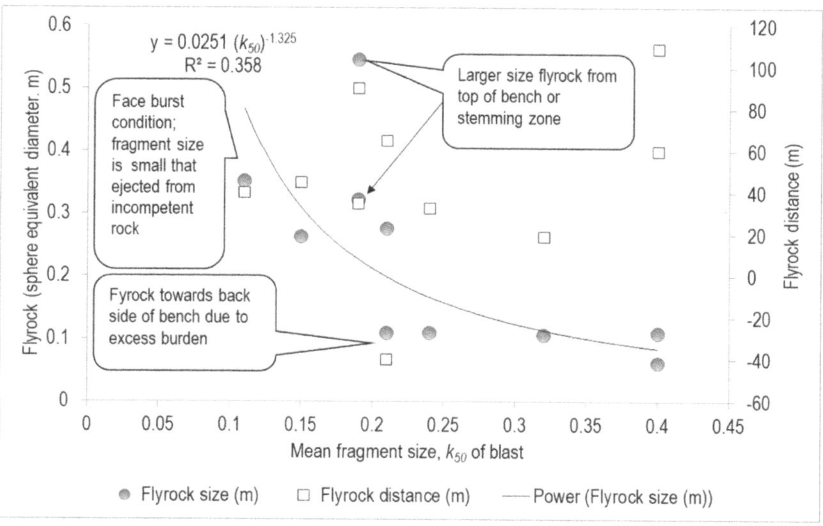

FIGURE 4.16 Relationship of flyrock size with the mean fragment size of blast and associated flyrock distance.

that will be ejected. This will require an exhaustive data base of flyrock size from all types of formations and drill diameter ranges that will provide a firm reason for predicting the flyrock size through correlations. However, it is sure that the flyrock size will not be larger than the largest block size formed by the joints in the rockmass being blasted and restricted by the pattern of (burden × spacing) in a blast. A complete understanding of the process with fragmentation size distribution vis-à-vis blast design is hence required to correlate the size of flyrock travelling a particular distance and the shape of the flyrock.

Trials in blasting in multiple rock and blasting conditions, with a limited data, were attempted by the author to correlate the flyrock size (k_f) with the mean fragment size obtained in such blasts (Figure 4.16). Despite the limited data and limitations of empirical analysis, the results show that there is an agreement between the mean fragment size of the blast and equivalent diameter of the flyrock recorded. Also, the flyrock distance in the case of larger fragment sizes is longer.

However, there is a discrepancy in the definition of the k_f owing to irregular nature of such rock fragments from blasting. It is not yet clear that which dimension of the flyrock should be considered for analysis. The criterion, whether the nominal diameter or the largest side should be considered, as the damage to a structure or injury to a person will depend on the exact position of the flyrock at the time of landing, needs to be established. If such conditions are imposed on predictions, the process becomes extremely complicated.

4.9 SIMULATION OF FLYROCK DISTANCE WITH AIR DRAG

A simulation of flyrock distance is carried out with the help of MATLAB© using app (Kim, 2022) that employs initial velocity, launch angle, mass of the fragment (flyrock in our case), drag coefficient, and slope of the impacting surface as inputs. The app yields the horizontal distance by using solution to non-linear differential equations without drag and later corrects the distances with the application of drag coefficient. The objective of the simulation was to test whether flyrock distance can be resolved with the use of the input variables.

Random sets of variables were created by autonomous functions for 50 tests. The values of initial velocity was 43–149 m/s, launch angle 16–83°, mass of the fragment 1–99 kg, drag coefficient 0.4–1.3, and slope of the surface 0.01–0.1. The relation of flyrock distance with the slope was not satisfactory as it is an extraneous factor. The slope of the surface was eliminated from the input variables and flyrock distances simulated using 19 regression learner algorithms out of which the Gaussian process regression (GPR, Figure 4.17) proved to be the best predictor yielding an excellent correlation with an R^2 of 0.96, RMSE of 13.7, and MAE 10.69. The test results are promising and can be improved with incorporation of other variables like wind velocity to evolve a simple method for flyrock travel distance.

4.10 THE FLYROCK DISTANCE PREDICTION: FOLLY OR TRUTH

It has been stressed and demonstrated earlier that flyrock is by chance and not design as also concluded by Richards and Moore (2004). The prediction of flyrock

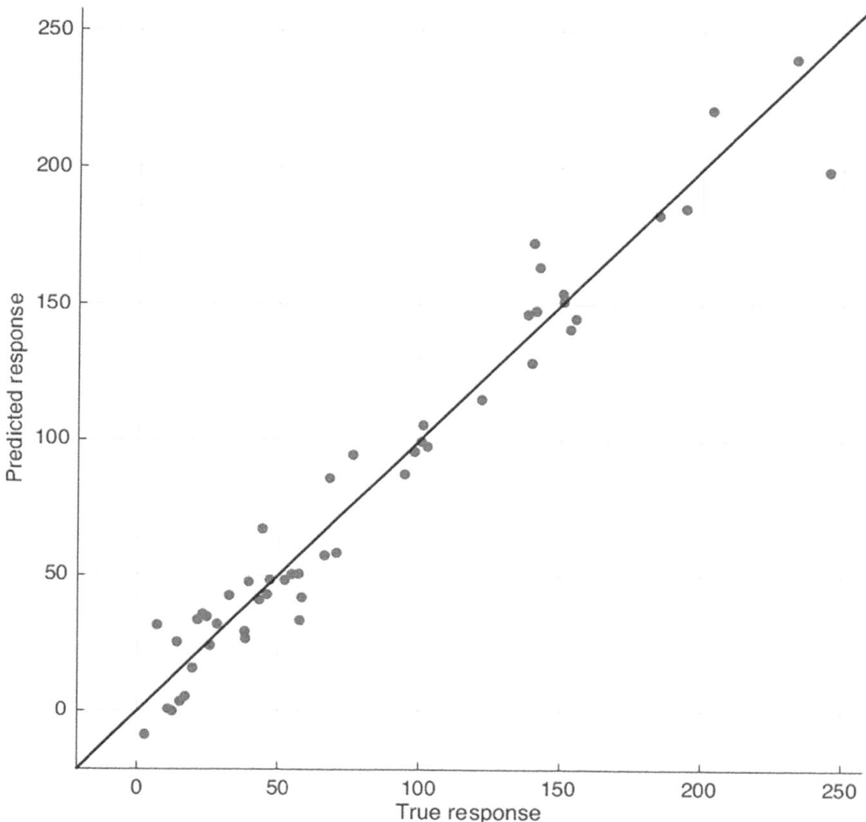

FIGURE 4.17 Predicted (made using ballistic equations with air drag) versus GPR estimated flyrock distance in metres.

is constrained by modelling of uncertain occurrence and not a regular phenomenon like throw and fragmentation. This situation leads to outlier probability estimation. In general statistics, such outliers could be easily accounted or pruned, but in the case of flyrock, these outliers are the concern of a blaster and hence cannot be eliminated. Combining such probabilities with flyrock distance predictions is a complex proposition. Defining the probabilities of flyrock events is a far more important consideration and should be an integral part of any prediction method. Also, there are several models and methods which favour prediction of flyrock distance, but these should be deployed with caution owing to non-differentiation between throw and flyrock, which in turn have different origins and mechanisms. While leaving the question of prediction of flyrock distance to the reader, it is practical to state that prevention is better than cure.

NOTE

1. www.lboro.ac.uk/microsites/research/phys-geog/tri-plot/index.html

REFERENCES

Aghajani-Bazzazi, A., Osanloo, M., & Azimi, Y. (2010). Flyrock prediction by multiple regression analysis in Esfordi phosphate mine of Iran. In *Rock fragmentation by blasting—proceedings of the 9th international symposium on rock fragmentation by blasting, FRAGBLAST 9* (pp. 649–657). CRC Press.

Amini, H., Gholami, R., Monjezi, M., Torabi, S. R., & Zadhesh, J. (2012). Evaluation of flyrock phenomenon due to blasting operation by support vector machine. *Neural Computing and Applications*, *21*(8), 2077–2085. https://doi.org/10.1007/s00521-011-0631-5

Armaghani, D. J., Hajihassani, M., Mohamad, E. T., Marto, A., & Noorani, S. A. (2014). Blasting-induced flyrock and ground vibration prediction through an expert artificial neural network based on particle swarm optimization. *Arabian Journal of Geosciences*, *7*(12), 5383–5396. https://doi.org/10.1007/s12517-013-1174-0

Armaghani, D. J., Koopialipoor, M., Bahri, M., Hasanipanah, M., & Tahir, M. M. (2020). A SVR-GWO technique to minimize flyrock distance resulting from blasting. *Bulletin of Engineering Geology and the Environment*, *79*(8), 4369–4385. https://doi.org/10.1007/s10064-020-01834-7

Armaghani, D. J., Mohamad, E. T., Hajihassani, M., Abad, S. A. N. K., Marto, A., & Moghaddam, M. R. (2016). Evaluation and prediction of flyrock resulting from blasting operations using empirical and computational methods. *Engineering with Computers*, *32*(1), 109–121. https://doi.org/10.1007/s00366-015-0402-5

Asl, P. F., Monjezi, M., Hamidi, J. K., & Armaghani, D. J. (2018). Optimization of flyrock and rock fragmentation in the Tajareh limestone mine using metaheuristics method of firefly algorithm. *Engineering with Computers*, *34*(2), 241–251. https://doi.org/10.1007/s00366-017-0535-9

Bakhtavar, E., Nourizadeh, H., & Sahebi, A. A. (2017). Toward predicting blast-induced flyrock: A hybrid dimensional analysis fuzzy inference system. *International Journal of Environmental Science and Technology*, *14*(4), 717–728. https://doi.org/10.1007/s13762-016-1192-z

Barkhordari, M. S., Armaghani, D. J., & Fakharian, P. (2022). Ensemble machine learning models for prediction of flyrock due to quarry blasting. *International Journal of Environmental Science and Technology*, *19*(6). https://doi.org/10.1007/s13762-022-04096-w

Berta, G. (1990). Explosives : An engineering tool. In M. Italesplosivi (Ed.), *Italesplosive* (Vol. 6). https://adams.marmot.org/Record/.b16176315

Bhatawdekar, R. M., Kumar, D., Armaghani, D. J., Mohamad, E. T., Roy, B., & Pham, B. T. (2020). A novel intelligent ELM-BBO technique for predicting distance of mine blasting-induced flyrock. *Natural Resources Research*, *29*(6), 4103–4120. https://doi.org/10.1007/s11053-020-09676-6

Bhatawdekar, R. M., Nguyen, H., Rostami, J., Bui, X. N., Armaghani, D. J., Ragam, P., & Mohamad, E. T. (2021). Prediction of flyrock distance induced by mine blasting using a novel Harris Hawks optimization-based multi-layer perceptron neural network. *Journal of Rock Mechanics and Geotechnical Engineering*, *13*(6), 1413–1427. https://doi.org/10.1016/j.jrmge.2021.08.005

Bieniawski, Z. T. (1989). Engineering rock mass classifications: A complete manual for engineers and geologists in mining, civil, and petroleum engineering. In *Engineering rock mass classifications: A complete manual for engineers and geologists in mining, civil, and petroleum engineering*. John Wiley & Sons.

Blanchier, A. (2013). Quantification of the levels of risk of flyrock. In *Rock fragmentation by blasting, FRAGBLAST 10—proceedings of the 10th international symposium on rock fragmentation by blasting* (pp. 549–553). CRC Press/Balkema. https://doi.org/10.1201/b13759-77

Calnan, J. T. (2015). Determination of explosive energy partition values in rock blasting through small-scale testing. *ProQuest Dissertations and Theses, 168*. https://search.proquest.com/docview/1764221237?accountid=188395

Chernigovskii, A. A. (1985). *Application of directional blasting in mining and civil engineering*. Oxidian Press India Private Ltd.

Chiappetta, R. F., Bauer, A., Dailey, P. J., & Burchell, S. L. (1983). Use of high-speed motion picture photography in blast evaluation and design. In *Proceedings of the annual conference on explosives and blasting technique* (pp. 258–309). International Society of Explosive Engineers.

Cunningham, C. V. (2006). Concepts of blast hole pressure applied to blast design. *Fragblast, 10*(1–2), 33–45. https://doi.org/10.1080/13855140600852977

Davies, P. A. (1995). Risk-based approach to setting of flyrock "danger zones" for blast sites." *Transactions—Institution of Mining & Metallurgy, Section A*, (May–August), 96–100. https://doi.org/10.1016/0148-9062(95)99212-g

Dehghani, H., Pourzafar, M., & Asadi zadeh, M. (2021). Prediction and minimization of blast-induced flyrock using gene expression programming and cuckoo optimization algorithm. *Environmental Earth Sciences, 80*(1). https://doi.org/10.1007/s12665-020-09300-z

Dehghani, H., & Shafaghi, M. (2017). Prediction of blast-induced flyrock using differential evolution algorithm. *Engineering with Computers, 33*(1), 149–158. https://doi.org/10.1007/s00366-016-0461-2

Duvall, W. I., & Atchison, T. C. (1957). *Rock breakage by explosives* (Report No. 5356). US Bureau of Mines, 52p.

Duvall, W. I., & Petkof, B. (1959). Spherical propagation of explosion-generated strain pulses in rock. In *Report of investigations/United States Department of the Interior, Bureau of Mines ;5483* (Issues ii, 21 p.). http://hdl.handle.net/2027/mdp.39015078542423

Faradonbeh, R. S., Armaghani, D. J., Amnieh, H. B., & Mohamad, E. T. (2018). Prediction and minimization of blast-induced flyrock using gene expression programming and firefly algorithm. *Neural Computing and Applications, 29*(6), 269–281. https://doi.org/10.1007/s00521-016-2537-8

Faradonbeh, R. S., Armaghani, D. J., & Monjezi, M. (2016). Development of a new model for predicting flyrock distance in quarry blasting: A genetic programming technique. *Bulletin of Engineering Geology and the Environment, 75*(3), 993–1006. https://doi.org/10.1007/s10064-016-0872-8

Faradonbeh, R. S., Armaghani, D. J., Monjezi, M., & Mohamad, E. T. (2016). Genetic programming and gene expression programming for flyrock assessment due to mine blasting. *International Journal of Rock Mechanics and Mining Sciences, 88*, 254–264. https://doi.org/10.1016/j.ijrmms.2016.07.028

Fattahi, H., & Hasanipanah, M. (2021). An integrated approach of ANFIS-grasshopper optimization algorithm to approximate flyrock distance in mine blasting. *Engineering with Computers, 38*, 2619–2631. https://doi.org/10.1007/s00366-020-01231-4

Flinchum, R., & Rapp, D. (1993). Reduction of air blast and flyrock. *Proceedings of International Society of Explosive Engineers, 147*.

Ghasemi, E., Amini, H., Ataei, M., & Khalokakaei, R. (2014). Application of artificial intelligence techniques for predicting the flyrock distance caused by blasting operation. *Arabian Journal of Geosciences, 7*(1), 193–202. https://doi.org/10.1007/s12517-012-0703-6

Ghasemi, E., Sari, M., & Ataei, M. (2012). Development of an empirical model for predicting the effects of controllable blasting parameters on flyrock distance in surface mines. *International Journal of Rock Mechanics and Mining Sciences, 52*, 163–170. https://doi.org/10.1016/j.ijrmms.2012.03.011

Graham, D. J., & Midgley, N. G. (2000). Graphical representation of particle shape using triangular diagrams: An Excel spreadsheet method. *Earth Surface Processes and Landforms*, *25*(13), 1473–1477.

Guo, H., Zhou, J., Koopialipoor, M., Armaghani, D. J., & Tahir, M. M. (2021). Deep neural network and whale optimization algorithm to assess flyrock induced by blasting. *Engineering with Computers*, *37*(1), 173–186. https://doi.org/10.1007/s00366-019-00816-y

Gupta, R. N. (1990). Surface blasting and its impact on environment. In R. K. Trivedy & M. P. Sinha (Eds.), *Impact of mining on environment* (pp. 23–24). Ashish Publishing House.

Gurney, R. W. (1943). *The initial velocities of fragments from bombs, shell, and grenades* (Report No. 405). Ballistic Research Laboratories, Aberdeen Proving Ground, 14p. https://apps.dtic.mil/sti/pdfs/ADA800105.pdf

Han, H., Armaghani, D. J., Tarinejad, R., Zhou, J., & Tahir, M. M. (2020). Random forest and bayesian network techniques for probabilistic prediction of flyrock induced by blasting in quarry sites. *Natural Resources Research*, *29*(2), 655–667. https://doi.org/10.1007/s11053-019-09611-4

Hasanipanah, M., & Amnieh, H. B. (2020). A fuzzy rule-based approach to address uncertainty in risk assessment and prediction of blast-induced flyrock in a quarry. *Natural Resources Research*, *29*(2), 669–689. https://doi.org/10.1007/s11053-020-09616-4

Hasanipanah, M., Faradonbeh, R. S., Armaghani, D. J., Amnieh, H. B., & Khandelwal, M. (2017a). Development of a precise model for prediction of blast-induced flyrock using regression tree technique. *Environmental Earth Sciences*, *76*(1), 27. https://doi.org/10.1007/s12665-016-6335-5

Hasanipanah, M., Jahed Armaghani, D., Bakhshandeh Amnieh, H., Koopialipoor, M., & Arab, H. (2018). A risk-based technique to analyze flyrock results through rock engineering system. *Geotechnical and Geological Engineering*, *36*(4), 2247–2260. https://doi.org/10.1007/s10706-018-0459-1

Hasanipanah, M., Jahed Armaghani, D., Bakhshandeh Amnieh, H., Majid, M. Z. A., & Tahir, M. M. D. (2017b). Application of PSO to develop a powerful equation for prediction of flyrock due to blasting. *Neural Computing and Applications*, *28*, 1043–1050. https://doi.org/10.1007/s00521-016-2434-1

Hasanipanah, M., Keshtegar, B., Thai, D. K., & Troung, N. T. (2022). An ANN-adaptive dynamical harmony search algorithm to approximate the flyrock resulting from blasting. *Engineering with Computers*, *38*(2), 1257–1269. https://doi.org/10.1007/s00366-020-01105-9

Holmberg, R. (1978). *Throw of flyrock in connection with rock blasting*. Technical Report. Swedish Detonic Foundation, 15p.

Holmberg, R., & Persson, G. (1976). *The effect of stemming on the distance of throw of flyrock in connection with hole diameters*. Report DS, 1. Swedish Detonic Research Foundation.

Hosseini, S., Poormirzaee, R., Hajihassani, M., & Kalatehjari, R. (2022). An ANN-fuzzy cognitive map-based Z-number theory to predict flyrock induced by blasting in open-pit mines. *Rock Mechanics and Rock Engineering*, *55*, 4373–4390. https://doi.org/10.1007/s00603-022-02866-z

Huandong, P., Congmou, L., & Jingyu, W. (2002). The dynamic process of flying stones in blasting and prevention measures against risk. In *Proceedings of the 7th international symposium on fragmentation by blasting, Fragblast 7* (pp. 703–707). Metallurgical Industry Press.

Hustrulid, W. (1999). *Blasting principles for open pit mining: Volume 1—general design concepts*. Balkema.

Jahed Armaghani, D., Tonnizam Mohamad, E., Hajihassani, M., Alavi Nezhad Khalil Abad, S. V., Marto, A., & Moghaddam, M. R. (2016). Evaluation and prediction of flyrock resulting from blasting operations using empirical and computational methods. *Engineering with Computers, 32*(1), 109–121. https://doi.org/10.1007/s00366-015-0402-5

Jamei, M., Hasanipanah, M., Karbasi, M., Ahmadianfar, I., & Taherifar, S. (2021). Prediction of flyrock induced by mine blasting using a novel kernel-based extreme learning machine. *Journal of Rock Mechanics and Geotechnical Engineering, 13*(6), 1438–1451. https://doi.org/10.1016/j.jrmge.2021.07.007

Kalaivaani, P. T., Akila, T., Tahir, M. M., Ahmed, M., & Surendar, A. (2020). A novel intelligent approach to simulate the blast-induced flyrock based on RFNN combined with PSO. *Engineering with Computers, 36*(2), 435–442. https://doi.org/10.1007/s00366-019-00707-2

Khandelwal, M., & Monjezi, M. (2013). Prediction of flyrock in open pit blasting operation using machine learning method. *International Journal of Mining Science and Technology, 23*(3), 313–316. https://doi.org/10.1016/j.ijmst.2013.05.005

Khanukayev, A. N. (1991). Perfection of blast works on the basis of new means and technology. In *Proceedings of the international conference of engineering blasting technique* (pp. 4–7). Peking University Press.

Kim, M. (2022). Projectile app. *MATLAB Central File Exchange*. www.mathworks.com/matlabcentral/fileexchange/52222-projectile-app

King, B. M., Just, G. D., & McKenzie, C. K. (1988). Improved evaluation concepts in blast design. *Transactions of the Institution of Mining and Metallurgy, Section A: Mining Technology, 97*, 173–181.

Konya, C. J. (1995). *Blast design* (1st ed.). Intercontinental Development Corporation. https://books.google.co.in/books?id=AdcOAQAAMAAJ

Koopialipoor, M., Fallah, A., Armaghani, D. J., Azizi, A., & Mohamad, E. T. (2019). Three hybrid intelligent models in estimating flyrock distance resulting from blasting. *Engineering with Computers, 35*(1), 243–256. https://doi.org/10.1007/s00366-018-0596-4

Ladegaard-Pedersen, A., & Persson, A. (1973). *Flyrock in blasting II, experimental investigation* (SweDeFo Report No. 13). Swedish Detonic Research Foundation.

Lemesh, N. I., & Pozdnyakov, B. V. (1972). The kinematics of motion of ledge rock in the zone of fracture during blasting. *Soviet Mining, 8*(4), 388–391.

Little, T. N. (2007). Flyrock risk. *Australasian Institute of Mining and Metallurgy Publication Series*, 35–43.

Little, T. N., & Blair, D. P. (2010). Mechanistic Monte Carlo models for analysis of flyrock risk. In *Rock fragmentation by blasting - proceedings of the 9th international symposium on rock fragmentation by blasting, FRAGBLAST 9* (pp. 641–647) CRC Press.

Lu, X., Hasanipanah, M., Brindhadevi, K., Amnieh, H. B., & Khalafi, S. (2020). ORELM: A novel machine learning approach for prediction of flyrock in mine blasting. *Natural Resources Research, 29*(2), 641–654. https://doi.org/10.1007/s11053-019-09532-2

Lundborg, N. (1974). *The hazards of flyrock in rock blasting* (SweDeFo Report No. DS 1974). Swedish Detonic Research Foundation.

Lundborg, N., Persson, A., Ladegaard-Pedersen, A., & Holmberg, R. (1975). Keeping the lid on flyrock in open-pit blasting. *Engineering and Mining Journal, 176*(5), 95–100. https://doi.org/10.1016/0148-9062(75)91215-2

Marto, A., Hajihassani, M., Jahed Armaghani, D., Tonnizam Mohamad, E., & Makhtar, A. M. (2014). A novel approach for blast-induced flyrock prediction based on imperialist competitive algorithm and artificial neural network. *Scientific World Journal, 2014*. https://doi.org/10.1155/2014/643715

McKenzie, C. (2009). Flyrock range and fragment size prediction. In *Proceedings of the 35th annual conference on explosives and blasting technique* (Vol. 2, p. 2). http://docs.isee.org/ISEE/Support/Proceed/General/09GENV2/09v206g.pdf

Monjezi, M., Amini Khoshalan, H., & Yazdian Varjani, A. (2012a). Prediction of flyrock and backbreak in open pit blasting operation: A neuro-genetic approach. *Arabian Journal of Geosciences*, *5*(3), 441–448. https://doi.org/10.1007/s12517-010-0185-3

Monjezi, M., Bahrami, A., Varjani, A. Y., & Sayadi, A. R. (2011). Prediction and controlling of flyrock in blasting operation using artificial neural network. *Arabian Journal of Geosciences*, *4*(3–4), 421–425. https://doi.org/10.1007/s12517-009-0091-8

Monjezi, M., Bahrami, A., & Yazdian Varjani, A. (2010). Simultaneous prediction of fragmentation and flyrock in blasting operation using artificial neural networks. *International Journal of Rock Mechanics and Mining Sciences*, *47*(3), 476–480. https://doi.org/10.1016/j.ijrmms.2009.09.008

Monjezi, M., Dehghani, H., Shakeri, J., & Mehrdanesh, A. (2021). Optimization of prediction of flyrock using linear multivariate regression (LMR) and gene expression programming (GEP)—Topal Novin mine, Iran. *Arabian Journal of Geosciences*, *14*(15). https://doi.org/10.1007/s12517-021-07772-2

Monjezi, M., Dehghani, H., Singh, T. N., Sayadi, A. R., & Gholinejad, A. (2012b). Application of TOPSIS method for selecting the most appropriate blast design. *Arabian Journal of Geosciences*, *5*(1), 95–101. https://doi.org/10.1007/s12517-010-0133-2

Ouchterlony, F., Nyberg, U., Mats, O., Ingvar, B., Lars, G., & Henrik, G. (2004). Where does the explosive energy in rock blasting rounds go? *Science and Technology of Energetic Materials*, *65*(2), 54–63.

Persson, P. A., Holmberg, R., & Lee, J. (1994). *Rock blasting and explosives engineering*. CRC Press. https://doi.org/10.5860/choice.31-5469

Rad, H. N., Bakhshayeshi, I., Wan Jusoh, W. A., Tahir, M. M., & Foong, L. K. (2020). Prediction of flyrock in mine blasting: A new computational intelligence approach. *Natural Resources Research*, *29*(2), 609–623. https://doi.org/10.1007/s11053-019-09464-x

Rad, H. N., Hasanipanah, M., Rezaei, M., & Eghlim, A. L. (2018). Developing a least squares support vector machine for estimating the blast-induced flyrock. *Engineering with Computers*, *34*(4), 709–717. https://doi.org/10.1007/s00366-017-0568-0

Raina, A. K., Chakraborty, A. K., Choudhury, P. B., & Sinha, A. (2011). Flyrock danger zone demarcation in opencast mines: A risk based approach. *Bulletin of Engineering Geology and the Environment*, *70*(1), 163–172. https://doi.org/10.1007/s10064-010-0298-7

Raina, A. K., Chakraborty, A. K., Ramulu, M., & Choudhury, P. B. (2006a). Design of factor of safety based criterion for control of flyrock/throw and optimum fragmentation. *Journal of the Institution of Engineers*, *87*, 13–17.

Raina, A. K., Chakraborty, A. K., Ramulu, M., & Choudhury, P. B. (2006b). *Flyrock prediction and control in opencast metal mines in India for safe deep-hole blasting near habitats—a futuristic approach* (Report). CSIR-Central Institute of Mining and Fuel Research, 156p.

Raina, A. K., & Murthy, V. M. S. R. (2016). Prediction of flyrock distance in open pit blasting using surface response analysis. *Geotechnical and Geological Engineering*, *34*(1), 15–28. https://doi.org/10.1007/s10706-015-9924-2

Raina, A. K., Murthy, V. M. S. R., & Soni, A. K. (2015a). Estimating flyrock distance in bench blasting through blast induced pressure measurements in rock. *International Journal of Rock Mechanics and Mining Sciences*, *76*, 209–216. https://doi.org/10.1016/j.ijrmms.2015.03.002

Raina, A. K., Murthy, V. M. S. R., & Soni, A. K. (2015b). Flyrock in surface mine blasting: Understanding the basics to develop a predictive regime. *Current Science*, *108*(4), 660–665. https://doi.org/10.18520/CS/V108/I4/660-665

Rezaei, M., Monjezi, M., & Yazdian Varjani, A. (2011). Development of a fuzzy model to predict flyrock in surface mining. *Safety Science*, *49*(2), 298–305. https://doi.org/10.1016/j.ssci.2010.09.004

Richards, A., & Moore, A. (2004). Flyrock control—by chance or design. In *Proceedings of the 30th annual conference on explosives and blasting technique* (pp. 335–348). International Society of Explosive Engineers.

Roth, J. A. (1979). *A model for the determination of flyrock range as a function of shot conditions* (Report No. PB81222358). US Department of Commerce, NTIS, 61p.

Rustan, A., Cunningham, C. V. B., Fourney, W., & Spathis, A. (2011). Mining and rock construction technology desk reference. In A. Rustan (Ed.), *Mining and rock construction technology desk reference*. CRC Press. https://doi.org/10.1201/b10543

Saghatforoush, A., Monjezi, M., Faradonbeh, & Armaghani, D. J. (2016). Combination of neural network and ant colony optimization algorithms for prediction and optimization of flyrock and back-break induced by blasting. *Engineering with Computers*, *32*(2), 255–266. https://doi.org/10.1007/s00366-015-0415-0

Sanchidrián, J. A., & Ouchterlony, F. (2017). A distribution-free description of fragmentation by blasting based on dimensional analysis. *Rock Mechanics and Rock Engineering*, *50*(4), 781–806. https://doi.org/10.1007/s00603-016-1131-9

Sanchidrián, J. A., Segarra, P., & López, L. M. (2007). Energy components in rock blasting. *International Journal of Rock Mechanics and Mining Sciences*, *44*(1), 130–147. https://doi.org/10.1016/j.ijrmms.2006.05.002

Shakeri, J., Khoshalan, H. A., Dehghani, H., M., B., & Onyelowe, K. (2022). Developing new models for flyrock distance assessment in open-pit mines. *Journal of Mining and Environment*, *15*. https://doi.org/10.22044/jme.2022.11805.2170

Sneed, E. D., & Folk, R. L. (1958). Pebbles in the lower Colorado River, Texas a study in particle morphogenesis. *The Journal of Geology*, *66*(2), 114–150.

St. George, J. D., & Gibson, M. F. L. (2001). Estimation of flyrock travel distances: A probabilistic approach. In *EXPLO 2001 conference* (pp. 409–415). AusIMM.

Stojadinović, S., Lilić, N., Obradović, I., Pantović, R., & Denić, M. (2016). Prediction of flyrock launch velocity using artificial neural networks. *Neural Computing and Applications*, *27*(2), 515–524. https://doi.org/10.1007/s00521-015-1872-5

Stojadinović, S., Pantović, R., & Žikić, M. (2011). Prediction of flyrock trajectories for forensic applications using ballistic flight equations. *International Journal of Rock Mechanics and Mining Sciences*, *48*(7), 1086–1094. https://doi.org/10.1016/j.ijrmms.2011.07.004

Stojadinović, S., Petrović, D., Ivaz, J., & Stojković, P. (2021). A Neuro-numeric approach for flyrock prediction and safe distances definition. *Mining, Metallurgy and Exploration*, *38*(6), 2453–2466. https://doi.org/10.1007/s42461-021-00512-w

Trivedi, R., Singh, T. N., & Gupta, N. (2015). Prediction of blast-induced flyrock in opencast mines using ANN and ANFIS. *Geotechnical and Geological Engineering*, *33*(4), 875–891. https://doi.org/10.1007/s10706-015-9869-5

Trivedi, R., Singh, T. N., & Raina, A. K. (2014). Prediction of blast-induced flyrock in Indian limestone mines using neural networks. *Journal of Rock Mechanics and Geotechnical Engineering*, *6*(5), 447–454. https://doi.org/10.1016/j.jrmge.2014.07.003

Trivedi, R., Singh, T. N., & Raina, A. K. (2016). Simultaneous prediction of blast-induced flyrock and fragmentation in opencast limestone mines using back propagation neural network. *International Journal of Mining and Mineral Engineering*, *7*(3), 237–252. https://doi.org/10.1504/IJMME.2016.078350

Udy, L. L., & Lownds, C. M. (1991). Partition of energy in blasting with non-ideal explosives. In *Proceedings of the annual conference on explosives and blasting technique* (pp. 193–194). International Society of Explosive Engineers. https://doi.org/10.1016/0148-9062(92)93963-k

Workman, J. L., & Calder, P. N. (1994). Predicting and controlling excessive flyrock. *Coal*, *99*(9), 26. https://doi.org/10.1016/0148-9062(95)94729-9

Wu, M., Cai, Q., & Shang, T. (2019). Assessing the suitability of imperialist competitive algorithm for the predicting aims: An engineering case. *Engineering with Computers*, *35*(2), 627–636. https://doi.org/10.1007/s00366-018-0621-7

Yari, M., Bagherpour, R., Jamali, S., & Shamsi, R. (2016). Development of a novel flyrock distance prediction model using BPNN for providing blasting operation safety. *Neural Computing and Applications*, *27*(3), 699–706. https://doi.org/10.1007/s00521-015-1889-9

Ye, J., Koopialipoor, M., Zhou, J., Armaghani, D. J., & He, X. (2021). A novel combination of tree-based modeling and Monte Carlo simulation for assessing risk levels of flyrock induced by mine blasting. *Natural Resources Research*, *30*(1), 225–243. https://doi.org/10.1007/s11053-020-09730-3

Zhou, J., Aghili, N., Ghaleini, E. N., Bui, D. T., Tahir, M. M., & Koopialipoor, M. (2020a). A Monte Carlo simulation approach for effective assessment of flyrock based on intelligent system of neural network. *Engineering with Computers*, *36*(2), 713–723. https://doi.org/10.1007/s00366-019-00726-z

Zhou, J., Koopialipoor, M., Bhatawdekar, R. M., Fatemi, S. A., Tahir, M. M., Armaghani, D. J., & Li, C. (2020b). Use of intelligent methods to design effective pattern parameters of mine blasting to minimize flyrock distance. *Natural Resources Research*, *29*(2), 625–639. https://doi.org/10.1007/s11053-019-09519-z

5 Blast Danger Zone

Mining, like other industrial operations, has its own drawbacks. There are instances of injuries and fatalities that have been observed in mining. A comprehensive account of such incidents in mining has been documented by Nowrouzi et al. (2017). Blasting as such has its own share in such accidents and as documented in preceding sections, flyrock is one such cause of accidents. There are evidences that people have been hit, injured, or killed by flyrock even at a distance of 1 km (Carlson & Eggerding, 2000) from the blast site. Proper engineering can, however, prevent the flyrock and control it to fall in a secured area to prevent damages to OCs.

Hence, the regulatory authorities have placed restraint on the blasting owing to the imminent danger due to flyrock, in terms of a restricted zone around blasting. The onus of any accident due to blasting through such rules is on the mine management. The restricted zone is called the blast danger zone (BDZ). The basic tenet of such rules is that people should not be present in the BDZ during blasting and any flyrock should not cross the defined or imposed limits of mine. Any violation can result in severe consequences.

Before discussing the blast danger zone, knowledge of some important terms is essential which enables to introspect the flyrock phenomenon in a holistic way, particularly in the context of prevailing geomining conditions or objects including habitats of any nature near to a blast.

5.1 DIFFICULT GEOMINING CONDITIONS

Geomining conditions vary from mine to mine and place to place. Surface, openpit or opencast mines present variation in their size and method of operations. In addition, the locale of the mines and presence of villages, structures, and other assets in proximity of mines restricts mining practices that can be classified as difficult conditions of mining. The location of OCs with respect to blast face is a key consideration when defining the BDZ. There are a host of variations associated with OCs like size of the mines, the type of the operation, the nearness of the habitats to the mine, the general psychosomatic response of residents nearby to blasting (Raina et al., 2004) and the presence of one or more habitat about the mine boundary.

Hence, factors that determine difficult geomining conditions, present individually or in combination, are as follows:

1. The presence of OCs in proximity, i.e. within few hundred metres with respect to of blasting area.
2. Presence of more than one habitat on all sides of the mine or blasting area.
3. The topography of the area that may be hilly in nature with habitats at toe or slope of the hills.

DOI: 10.1201/9781003327653-5

4. The mine configurations, including limited working area due to space restrictions and hence throw restrictions, posing constraints on regular relief and confined of blastholes with adequate delays.
5. The type of rock excavated, like hard fractured and/or adversely dipping rock.
6. Presence of geological features like incompetent lithology in highly competent rocks, in situ boulders, voids and fissures, mixed face conditions, and natural features that can be detrimental during blasting.
7. Increased frequency of blasting due to space limitations, working conditions, and hence increased risk due to flyrock. This frequently results in choked blasts owing to non-removal of previously blasted muck.
8. Remote location of mine hampering the availability of proper explosives and accessories due to poor connectivity or security issues related to explosive transportation.

The definition of difficult geomining conditions can be used for ascertaining the blast design and defining the BDZ, while considering that such difficult geomining scenarios are highly prone to flyrock occurrence. The information can be vital in evaluating the risk due to flyrock and hence aid in optimizing the blast design and BDZ, through proper evaluation.

5.2 OBJECTS OF CONCERN IN MINING

Generally, there are two types of objects in an around mines. These may include objects that belong to the mines and those that do not belong to the mines. Objects within the mines and belonging to the mines can get damaged due to mining operations and flyrock and have a direct, yet manageable, consequence for the owner or operator of the mines. However, the case with the objects that are beyond the mines and are not owned by the owner of the mines have a different relationship with mines. Any damage to such objects due to any mining operation can invite significant penalties, including closure of the mines.

The presence of people within and outside the mines is a difficult situation, as flyrock can result in fatalities. Hence, it is important to have a holistic view of objects that can be in and around the mines and are amenable to flyrock. Such objects have been called "Objects of Concern" or OCs (Raina, 2014). The types of OCs, their nature, and consequences on the event of a flyrock impacting them are detailed in Table 5.1.

The interactions between the mines and a combination of aforementioned OCs present an intricate situation for the blasting engineer. Moreover, the cost of an event is relative in nature as the value of currency changes with time and it also varies from place to place. Although there are published reports with quantified penalties for flyrock occurrence defined in different countries, it is impossible to refer to these, compile, and present an absolute cost framework of such events. An alternative therefore will be to have a list of insurance liabilities in such cases. However, insights into the subject can be gained from the thesis (Loeb, 2012)[1] that provides a comprehensive account of regulations of the United States and Australia related to flyrock, particularly while blasting in urban areas, and a detailed account of flyrock incidents and the penalties thereof.

TABLE 5.1
Definition of Objects of Concern With Respect To Flyrock Risk Domain

S. No.	Objects of Concern (OCs)	Consequence	Mobility	Relative Cost of a Flyrock Event
1	Persons outside the mine	Injury, fatalities	Static	High
2	Persons within the mine or blasting zone	Injury or fatalities	Shifting	High
3	Structures beyond blast zone/ leasehold	Damage	Static	High
4	Structures within blast zone/ leasehold	Damage	Shifting	Low
5	Equipment within blast zone/ leasehold	Damage	Shifting	Low
6	Equipment outside blast zone/ leasehold	Damage	Static	Moderate
7	Livestock outside blast zone/ leasehold	Injury, fatality	Static	Moderate

Source: Modified after Raina (2014).

The orientation of the bench or blast free face with respect to the OCs has a significant relationship with flyrock. The directional probabilities of flyrock with respect to blast face was investigated through tests on concrete models, as discussed in Chapter 4, Section 4.8.1. The results of such tests indicated that the maximum likelihood of the throw of material and fragments is towards the front of the free face (Figure 5.1) and is in tune with the findings of Lundborg et al. (1975), Lundborg (1979), and (Richards & Moore, 2004).

The probability of a flyrock, moving in a particular direction in a mine, can thus be worked out in a similar way in actual field conditions and used to estimate the impact frequency of total probabilities associated with flyrock. The distance of OCs with respect to a blast face can also be determined in the direction of maximum probability of flyrock and used to determine the threat level and risk. The consequences and levels of severity of a flyrock travelling to a distance that can be considered dangerous for the nearby OCs can also be worked for a particular mining condition.

5.3 ACCIDENTS: THE REAL CONCERN

Verakis and Lobb (2003) compiled a statistics of 1139 incidents during the period 1978–2003, out of which they reported that blast area security accounted for 50.1% of these injuries followed by premature blast (11.4%), flyrock (10.8%), misfires (9.9%), and fumes (8.5%). A compilation of different such works with possible overlap in data as provided in Table 5.2 is an indication of accidents that happen due to blasting, with a major share from the flyrock or violation of the blast danger zone (BDZ). The latest data has not been updated by authors reporting flyrock, though

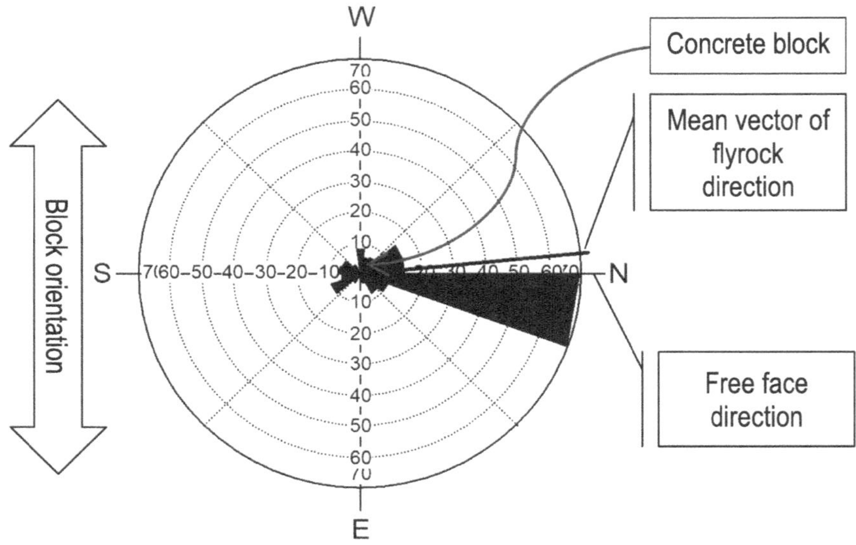

FIGURE 5.1 Direction of throw of blasted rock and flyrock from concrete model studies.

TABLE 5.2
Statistics of Accidents Related to Different Aspects of Blasting, Including Blast Area Security and Flyrock Project beyond the Control Area or Blast Danger Zone

Citation	Period	Blast Area Security	Flyrock Projected beyond Blast Area	All Blasting Injuries	% Flyrock Injuries
Bajpayee et al. (2004)	1978–1981	51	26	77	33.77
	1982–1985	28	22	50	44.00
	1986–1989	43	29	72	40.28
	1990–1993	25	24	49	48.98
	1994–1997	17	10	27	37.04
	1998	3	3	6	50.00
Verakis (2011)	2006–2010		18		38.00
Zhou et al. (2002)	–				27.00
Adhikari (1999)					20.00
Verakis and Lobb (2007)	1994–2005	68	32	168	19.05
Little (2007)	1978–2098	281	281	412	68.20
Kecojevic and Radomsky (2005)	1978–2001	89	54	195	27.69
Mishra and Mallick (2013)	1996–2011			30	24.19

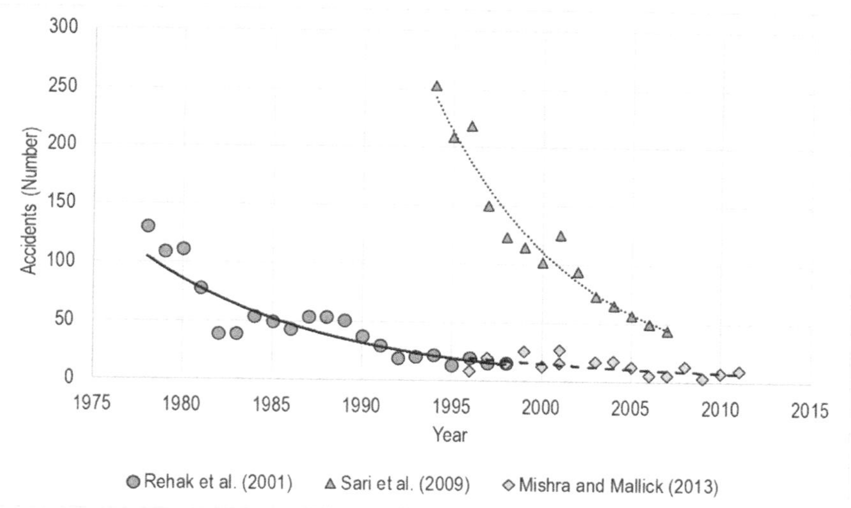

FIGURE 5.2 Accidents due to blasting/underground workings over a period of time.

there are significant number of reports predicting flyrock distance through observations. The reasons for the same are obvious and treated earlier also.

The general trend of accidents due to blasting in surface mines has seen a decrease over time as per the statistics (Rehak et al., 2001) for the period 1978–1998 and as visualized in Figure 5.2. The decreasing trend of such accident data (Mishra & Mallick, 2013) for the period conforms to the trend of the data of Rehak et al. (2001). This is also the case with the overall accident trend (Figure 5.2) from underground reported by Sari et al. (2009). These are evidence of improvement of safety standards owing to several reasons that can be related to mechanization, proper training, or even risk evaluation and economic appreciation, legislation and guidelines of accidents due to such activities.

The accident statistics, however, suggests that prediction of flyrock distance, that is universally acceptable, is still a need and the definition of a BDZ is required in scientific terms. A host of probabilities enter into the domain of such definition of BDZ as defined by Davies (1995). The ratio of flyrock injuries to the number of blasts conducted in a mine may change according to the blasting practice and appreciation of the concern towards flyrock by a particular mine and its management. This means that the BDZ should vary from mine to mine with some general considerations like orientation of the blast face and presence of OCs in the vicinity of the mine or blast.

Despite the aforementioned observations, the accidents still occur. Blasting has quite significant input to accident statistics and flyrock is one of the reasons of such accidents. The situation leads to several questions about the occurrence or probabilities with their changing definition, the distance, the size and direction of flyrock, and securing the blast area. Some of the questions have already been addressed in Chapter 4, but what should be the basis of defining the blast danger zone (BDZ) remains to be seen. Should it be, as accepted generally, i.e. just event-based, or there should be robust science behind defining the BDZ. This means a holistic approach to the flyrock that, in addition to the aforementioned questions, should cater to the impact of flyrock and cost analysis of such events.

So, the existing BDZ definitions, the legal framework, understanding of the requirements for defining a scientific BDZ, and how risk and other methods can be put to work for a logical resolution of the BDZ are discussed here.

5.4 BLAST DANGER ZONE: DEFINITION AND TERMINOLOGY

The incidences of flyrock from blast faces have resulted in formulation of blast danger zone or Blast Security Area around the blast sites. There are quite a few definitions in the published reports and regulations that refer to the blast danger zone (BDZ) and by different names. The terminology for the BDZ used are as follows:

1. Blast danger zone
2. Blast security zone
3. Blast area security
4. Blast area
5. Blast exclusion zone

However, all the terms mean the same thing. These represent the area around a proposed blast which should be cleared of people and equipment before conducting the blast to prevent accidents due to blasting and, more specifically, flyrock. Since the term BDZ incorporates a complete description, the same is proposed to be recognized and used in place of multiple terminologies. It is relevant to note that the lack of standardization of the terminology is to disadvantage of scientists, planners, field engineers, and legal authorities, as it broadens the search and the search criteria, and impedes the actual inputs required from them owing to ambiguity in definitions.

5.4.1 Legal Framework

Flyrock occurrences have resulted in several event-based definitions of blast danger zone about the blast in different countries. There are instances when BDZ has been changed several times through laws, rules, and regulations based on the flyrock travel distance of a single event. This has led to a same BDZ for all the mines, irrespective of their size, blasting practice, and execution culture, thus limiting production. The USBM standards declare that the danger zone should be half (½) of the distance of the objects of concern from the blast site. In India, Directorate General of Mines Safety (DGMS, 2003) has fixed the blast danger zone at 500 m, irrespective of the distance of the objects of concern from the blast site and mining conditions. This raises a doubt about the efficacy of such legal frameworks.

Arthur (1979) mentioned that restrictions against blasting apply where dwellings, schools, churches, hospitals, or nursing facilities are within 305 m (1000 ft) of blasting operations. Such rules also apply to presence of wells, petroleum or gas storage facilities, municipal water storage facilities, fluid transmission pipelines, gas or oil collection lines, and sewage lines, within 152 m (500 ft) of blasting operations. Dick (1979), while reviewing the US federal surface coal mine blasting regulations, mentioned that flyrock is the most potentially serious hazard to life and property. These regulations restrict the operator from throwing rock beyond his property limits or more than half the distance from the blast to any dwelling, public building, school, church, and commercial or institutional building. Also, flyrock must not be thrown beyond regulated BDZ. This, of course, does not apply to structures owned by the mine operator and not leased to other persons. A comparison of some statutes pertaining to flyrock has been presented by Rehak et al. (2001) and Bajpayee et al. (2003).

Most of the regulation world over are intertwined with explosive rules or mining buffer zones and may or may not be explicit with respect to flyrock AEISG (2011). The access to many of these is restricted or specific to their respective websites and makes search and acquisition somewhat difficult. Hence, it is not fit to claim to have a comprehensive account of the legislations and laws pertaining to flyrock. However, the most cited and analysed one's are summarized here.

International Labour office (ILO), Geneva, Code of Practice, Safety and Health in Opencast Mines, Under Rule 604, General Provisions 9.28.1 of Chapter Shot Firing, states that "*National laws and site procedures should specify the extent of the danger zone surrounding surface blasting operations, procedures to protect persons and property at risk of being affected by shock waves (vibration), flying fragments, dust and fumes from blasting operations.*"[2]

Newfoundland and Labrador, Canada, Quarry Materials Act (1998), provided that "*the minimum buffer distance of pit and quarry workings from existing or proposed residential development should be 300 m where no blasting is involved and 1000 m where blasting is involved.*"[3]

Chapter "Control of Adverse Effects" of the Office of Surface Mining and Reclamation and Enforcement (OSMRE), Department of Interior, US, under CFR 30 (section 816.67(c)) states:

> *Flyrock travelling in the air or along the ground shall not be cast from the blasting site, (1) More than one-half the distance to the nearest dwelling or other occupied structure; (2) Beyond the area of control required under section 816.66(c); or (3) Beyond the permit boundary (CFR 30, 701.5).*

Government of India, Regulation 164, 1-A(b) of the Metalliferous Mines Regulations, 1961 stated:

> *[I]n the case of an opencast working the blaster shall not charge or fire a shot, unless sufficient warning, by efficient signals or other means approved by the manager, is given over the entire area falling within a radius of 300 meters from the place of firing (hereinafter referred to as the danger zone) and also, he has ensured that all persons within such area have taken proper shelter.*[4]

However, the draft notification of the Gazette of India, 2020, amended Metalliferous Mines Regulations 1961 Act, section 2(b) of regulation 188 (under heading Taking Shelter), states:

[I]n the case of an opencast working, the blaster shall not charge or fire a shot, unless sufficient warning, by efficient signals or other means approved by the manager, is given over the entire area falling within a radius of 500 meters from the place of firing (hereinafter referred to as the danger zone) and also he has ensured that all persons within such area have taken proper shelter.[5]

There are instances of restriction of flyrock by Director General of Mines Safety (DGMS), India to 10 m, that state, "The manager of the mine shall ensure that flyrock shall not extend beyond 10 m from the place of firing."[6]

Department of Environment and Labour, Nova Scotia, Pit and Quarry guidelines, under Section 2(c) of rule IV (Separation Distance for Quarry Operations), 2003, states:

[N]o person responsible for the operation of a quarry shall blast within 800 m of the foundation or base of a structure located off site. Structure includes but is not limited to a private home, a cottage, an apartment building, a school, a church, a commercial building, a treatment facility associated with the treatment of municipal sewage, industrial or landfill effluent, an industrial building or structure, a hospital, nursing home etc., the separation distance is measured from the working face and point of blast to the foundation or base of the structure. This distance can be reduced with written consent from all individuals owning structures within 800 m.[7]

There are citations (Schwengler et al., 2007) of Australian/New Zealand Standard 2187.2–2006, Appendix E, that highlight blast design controls to minimize the generation of flyrock.

Bajpayee et al. (2002), while giving a comprehensive report of some erstwhile international regulations pertaining to flyrock, refers to an Australian document (DR 04062, 2004) issued for public comments that emphasizes the risks of blasting, and recommends the following:

[W]henever explosives are to be used that a competent person(s) carry out a detailed risk assessment to identify all foreseeable potential hazards and take appropriate steps to eliminate or reduce the likelihood and mitigate the severity of any effects so that risks are at an acceptable level.

The Australian document highlights Brnich and Mallett's (2003) hazard-risk assessment concept.

State Regulations, South Carolina Code of Regulations, Chapter 89—Office of The Governor-Mining Council of South Carolina S.C. Code Regs. 89–150—Surface Blasting Requirements, Clause I, states:

[T]o provide for adequate public safety, the operator shall be required to maintain a minimum distance between the nearest point of blasting and any structures not owned by the operator as of the completed application date or where there is no waiver of damage. The minimum distance shall be established by the Department after considering the method of mining, site conditions, proposed

directions of blasting, type and use of neighbouring structures, previous blasting record, and/or other factors as deemed appropriate by the Department.[8]

MHSC (Mine Health and Safety Act, 1996 (Act No. 29 of 1996) and Regulations, 1998), Clause 4.7 regulates:

[T]he employer at any mine must take reasonable measures to ensure that when blasting takes place, air and ground vibrations, shock waves and fly material (flyrock) are limited to such an extent and at such a distance from any building, public thoroughfare, railway, power line or any place where persons congregate to ensure that there is no significant risk to the health or safety of persons.

The interesting thing in the aforementioned statutes is that the stemming of blastholes is important (Clause 4.14(1)), and the length of the stemming and tamping is based on risk analysis done by the management in consultation with explosive manufacturer (Clause 4.14(2)) to contain flyrock. Clause 4.16(1)(a) emphasizes precautions in securing the safety of people, that is a direct indication of establishing a BDZ.

Some and not all the laws cited about BDZ lay the onus of flyrock on the mine management and is a better one that gives leverage to the mines to decide the BDZ and risk management. Few of such laws are quite explicit in terms of defining the risk by the mines and owning the consequences of the blasting. However, the universal code should focus on defining the BDZ on logic and risk involved in blasting. The fact that blasting is even permitted within BDZ through special permissions under the premise of controlled blasting justifies deploying a scientific basis for the zone.

5.4.2 Pre-requisites to Define Blast Danger Zone

The BDZs presented earlier do not discriminate between nature, type, blasting culture, and probability of the accidents in a mine, nature of OCs, and many other aspects of flyrock as brought out earlier. This constrains the mines in particular and the mining industry in general. Such influence of the regulations can manifest in less productive mining and unnecessary stressful conditions owing to extreme legislations.

The criteria of BDZ are somewhat vague in nature as it is generally marked as a circular zone on mine plans about the centre of the mines. The distance from the periphery to the blast will vary on daily basis and may increase or decrease, accordingly. Even if the BDZ is used in relation to a blast, it is difficult to mark the same and have a full, yet fool proof, control on the security zone. Also, mines are assumed to be of circular shape while as the shape in actual mining is rarely so. Most of the mines are elongate in shape, which have not been considered in BDZ. The case of linear mines, particularly large opencast coal mines, that extend in length over few thousand metres have an altogether different domain of OCs.

Hence, to define the blast danger zone in scientific terms, a summary of several pre-requisites, most of which have been addressed earlier, is presented here:

1. To define the occurrence of flyrock, i.e. as to what should be classified as flyrock. This means a clear discrimination between throw and flyrock.

2. What size flyrock can be thrown beyond a particular distance and what can be its damage potential.
3. The exit or maximum velocity, size, and launch angle of flyrock being thrown that depend on the rock, explosive, and blast design variables. This should finally define the flyrock distance.
4. Travel of the fragment under the influence of air drag, its final landing location, including influence of topography of a mine, rebound, and the Magnus effect.
5. The probability of the flyrock occurrence vis-à-vis topography of the mine, the orientation of dominant rock joints, and benches with respect to the objects of concern, and frequency of blasting.
6. Influence of various blast design variables and rockmass characteristics on flyrock distance.
7. The consequences of a flyrock hitting OCs. A detailed account of penalties on mine management through case studies and examples available world over can form its basis while drawing queue from the work of Loeb (2012).
8. The risk associated with flyrock if it hits an object of concern. Risk is a product of probability and consequence. It should be noted that the cost of a flyrock event is very difficult to work out in the case of fatalities.

This brings us to the first question as to what should be classified as flyrock? Rock fragments from a blast tend to project and even after fall these continue to cover some distance, as demonstrated in Chapter 2, that is not under control of a blasting engineer. Hence, any fragment and every fragment, that moved away from the face, cannot be classified as flyrock. A simple representation of individual fragments that were cast at different distances from a database of 367 blasts is presented in Figure 5.3, where it is observed that as the distance from the blast and OCs increases, the number

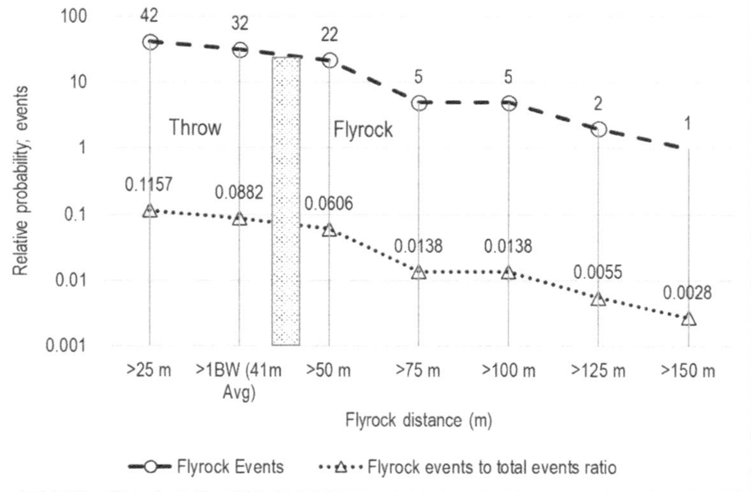

FIGURE 5.3 The relative probability and flyrock events considering varying distance of travel of a fragment as flyrock.

of fragments thrown a particular distance decrease. The analysis points to the fact that there is a need to classify a fragment as flyrock in terms of distance from the blast face. The bench configurations present a case, i.e. a fragment can be classified as flyrock if it moves couple of bench widths (see complete description and explanation given in Section 4.8.2). Bench width is proposed as a measure as fragments moving beyond such distances can be difficult to observe or may pose difficulty in mining process. Although this method presents a static classification scheme for a mine, but can be adopted universally, as shown by the shaded portion in Figure 5.3. However, a dynamic method of classification will be to define a zone around the blast and fragments that are thrown beyond the zone can be classified as flyrock. However, this method has several issues with definition, analysis, and reporting.

The objective of outlining the classification is that a universal, yet specific method will be available for researchers, mine owners, and legislators to place flyrock on equal footing, so far as its definition is concerned. This will facilitate the true reporting of flyrock, provide a basis for scientists to define the probabilities of flyrock, and ultimately define the risk due to flyrock in a universal domain.

The second question is about the size of the flyrock. As defined earlier, the maximum flyrock size cannot be more than the maximum size of the in situ block size. Also, larger fragments owing to larger area of impact by explosive gases have tendency to travel larger distances. So flyrock of small sizes can be ignored, if these are encountered in a blast. There are instances where fragments more than 0.1 m diameter have been considered as flyrock and this can form a basis for further classification of flyrock. By following the aforementioned criterion, if fragments of <0.1 m size are detected beyond a few bench widths, these can be ignored, unless these have travelled more than half the distance of the nearby OCs.

The calculation of exit velocity can be estimated from pressure–time relationships or by using empirical methods defined in Chapter 4. Travel of flyrock and its distance from the blast face can thus be estimated while deploying the air drag and other effects. The adjustments for landing impact can be accounted accordingly. A method provided in Chapter 4 for determining the level of safety along with the use of correction factor for flyrock in throw prediction can be used as an alternative.

5.5 RISK-BASED BLAST DANGER ZONE FOR FLYROCK

A risk-based BDZ for flyrock has three components, viz. the probability of flyrock crossing a particular distance from the blast zone, the consequences of such event in case the flyrock hits an object of concern, and how the risk is to be managed.

We will try to address the former two elements as these define the risk. The risk management will be taken up in the Chapter 6. Based on the findings cited earlier, the following approaches emerged for flyrock that can be used to define the blast danger zone:

1. Empirical method
2. Stochastic modelling of flyrock distance and estimation of the probability of flyrock. This involves calculation of the risk or defining the risk matrix in terms of the probability and consequences of a flyrock event.
3. Semi-empirical pressure–time-based criterion

5.5.1 Empirical Method

This is a simple method in which the maximum flyrock distance can be measured in a particular mining condition and the BDZ fixed accordingly. This is an observational method as is defined in most of the regulations and rules world over. The operator of a mine or the regulatory authorities can choose a factor of safety over the maximum flyrock distance that can assume a value of 1.5–2.0, depending upon the variability of flyrock distance and presence of vulnerable OCs in the vicinity of the mine. The method has a disadvantage that the probabilities and risk of failure of the safety rule are not considered as an option. Further the BDZ is a circular zone around the blast face.

5.5.2 Stochastic Method

The stochastic method involves data collection, defining distribution of the events or accidents, defining the severity of the events, and then simulating the results using Monte Carlo or other methods. The probabilities defined in this way along with the consequences can be used to predict risk associated with the event and finally making decisions about the BDZ and hence the blast design (Kecojevic & Radomsky, 2005).

Bandyopadhyay et al. (2003) discussed the potential application of fuzzy set theory in evaluating risk using linguistic variables/values. They (Bandyopadhyay et al., 2003) used the Yager's methodology for ordinal multiobjective decision based on fuzzy sets to evaluate risk due to environmental factors of blasting in mines, including flyrock. Little (2007) defined the consequence and risk matrix-based approaches for flyrock risk assessment while simulating conditions as per the criterion of Richards and Moore (2004). Little and Blair (2010) dwelled further on their previous work and devised methods for defining the mechanistic method of assessment of flyrock risk. They also brought out the pros and cons of the existing criteria of risk due to flyrock. It may be pointed out that the area of influence of flyrock and risk are quite related. The estimation of such phenomenon for inclusion will need enormous amount of data to quantify. Blanchier (2013) evaluated the statistical distribution of flyrock distances that takes into account rockmass and blast pattern and their variations.

The recent analysis of the flyrock occurrence (Blair, 2022) provided a distance (R_f) model based on the 2D flight characteristics of a projectile with site-specific blast design of different quarries. This 2D model shows that good correlation to the probabilities of measured R_f are consistent to known predicted ranges while describing the flyrock movement in radial and vertical directions. The author (Blair, 2022) also attempted a pseudo 3D model for R_f prediction, rotated randomly in radial direction around each blasthole to predict flyrock trajectories. However, the model does not cover all the random trajectories of the 3D surface and provides the results of projectile landing only along the chosen random paths of the model with a restrictive limitation owing to the vast number of flyrock influencing factors. The author tried to overcome this problem by generating large number of flyrock trajectories in the model, i.e. using 24.28 million radial directions for each of the four blastholes constructed in the model, which apparently gives 18.11 billion trajectories for pseudo 3D model and landing coordinates of the flyrock. However, these four blastholes are not interacting with each other, which raises the question for reliability of the model.

So, to define the blast danger zone, it is essential to work out the probabilities of an event and its consequences, which define the risk involved with the help of the method given in Equation 5.1.

$$\begin{gathered}\text{Risk} = \text{Probability of an event } P(E)\times \text{ Consequence of an event } C(E)\\ \text{Or}\\ \text{Risk } = P(E)\times C(E)\end{gathered} \tag{5.1}$$

Probability of a flyrock distance exceeding the permissible limit, $P(E)$, at a particular mine, can be worked out from the probability density function evolved over observed data. The probability of a flyrock event needs sufficient data and should be worked out by a mine while outlining a safety rule. It is important to mention that a distinction should be made in the probabilities of throw of the blasted material and those of the flyrock as these show different distributions. While throw has a normal distribution, the flyrock events show a logarithmic distribution that is highly skewed towards the lower values. However, the consequence in terms of cost and/or penalties arising from the flyrock event is not known or is difficult to estimate particularly in the case of fatalities, a Threat ratio (T_r) representing the consequences of a flyrock event (Raina et al., 2011), $C(E)$, that replaces the consequence that can be worked out with the help of Equation 5.2:

$$C(E) = T_r = \frac{R_{perm}}{R_{obj}} \tag{5.2}$$

where R_{perm} is the permissible or acceptable distance of flyrock in m, R_{obj} is the distance of object of concern from the blast site in m, and T_r is the threat ratio.

Lundborg (1979) calculated that at a distance of 600 m, while blasting a 76-mm diameter hole, the probability of being struck by a flyrock was the same as that being struck by a lightening (1 in 10,000,000). However, this probability may not hold good for cases where blasting is carried out in proximity of habitats. Lundborg (1979) further demonstrated that the probability of the flyrock distance assumes a Weibull distribution.

Moreover, flyrock probability has several subcomponents as enumerated by Davies (1995), who provided an approach to the setting of "Danger zones" by considering the incidence of flyrock, which is calculated from the available data, and the probability that a predicted distance will be exceeded. Davies (1995) also utilized the formula established by the Roth (1979) for calculation of flyrock distance and its exit velocity. He has explained the target impact frequency for both the individual and the societal or group risks of flyrock impact.

Normally, short distance flyrock travels less than 300 m. However, when there is some abnormality in a blast or rock formation, flyrock has been known to travel much further than the calculated distances. The frequency of impact by such flyrock at a constant distance, for single shot provided by Davies (1995), is given in Equation 5.3:

$$f = N_b \times f_f \times P(R) \times P(T) \times P(T_e) \tag{5.3}$$

TABLE 5.3
Probabilities of Flyrock

Location	Incidents/m³	Incidents/kg	Incidents/blast
1. United Kingdom			
A. Blasting quarries/mines	3.59×10^{-7}	1.41×10^{-10}	1.3×10^{-3}
B. Hard stone quarry	9.45×10^{-7}	3.64×10^{-10}	
2. Hong Kong			1.02×10^{-3}
Blasting quarries/mines	5.30×10^{-7}	2.0×10^{-10}	

Source: After Davies (1995).

where f is the target impact frequency (impact/year), N_b is the total number of blasts per year, f_f is the frequency of flyrock per blast, $P(R)$ is the probability of wild flyrock travelling the target distance, $P(T)$ is the probability of wild flyrock travelling on an impact trajectory, and $P(T_e)$ is the probability of target exposure.

The probabilities for different operations provided by Davis (1995) are given in Table 5.3, who advocated that owing to underreporting of flyrock incidents, the risk figures calculated earlier should be increased by a factor of 2–3. The probabilities are however presented here and can be used as baseline for evaluating risk.

The risk analysis method given by Davies (1995), although complex in nature, is a comprehensive one and worthy of consideration when significant economics of the operation is involved. The adoption of this criterion can be possible should the mines be ready to document the flyrock in each blast over a period. The main disadvantage of this method is that it is difficult to obtain realistic values for each component (Little & Blair, 2010).

Blanchier (2013) described that R_f for a blast in terms of probability p is $R_f = f(p)$ and calculated the impact zone of flyrock with a given probability as follows:

- For flyrock from a face

$$\Delta S = \pi d \Delta R_f = \pi f(p) \times f'(p) \times \Delta p$$

$$\Delta S \sim \pi f(p) \times \left[f\left(p + \frac{\Delta p}{2}\right) - f\left(p - \frac{\Delta p}{2}\right) \right]$$

- For flyrock from a blasting surface

$$\Delta S = 2\pi d \Delta R_f = 2\pi f(p) \times f'(p) \times \Delta p$$

$$\Delta S \sim 2\pi f(p) \times \left[f\left(p + \frac{\Delta p}{2}\right) - f\left(p - \frac{\Delta p}{2}\right) \right]$$

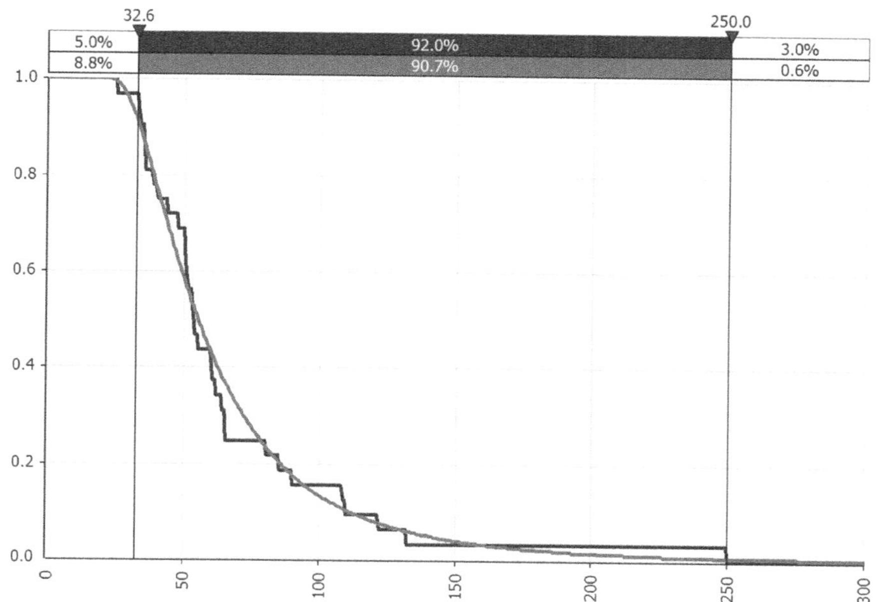

FIGURE 5.4 Cumulative probability plot of flyrock event with respect to travel distance (field blasts).

Considering the incidence angles and visible surface S_a of a person in the region of 0.1 m^2, he defined the probability p_a of a person being hit by a flyrock as in Equation 5.4:

$$p_a = \frac{\Delta p}{\Delta S} S_a \tag{5.4}$$

Further, counting on the number N of blasts per year in the direction the annual probability of impact $p_{a,i}$ (from the working face, considering front row holes only) is established through Equation 5.5 and can be finalized by an estimation of the probability of serious or fatal accidents to define the danger level for people (as all impacts in the aforementioned study are assumed to be fatal).

$$p_{a,i} = p_a N \tag{5.5}$$

where p is the level of probability, p_a is the probability of a person impacted by flyrock, S_a is the visible surface of a person (note this surface is in the region of 0.1 m^2, considering the incident angles), and N is the number of blasts per year.

To test the probability density function, the field blast data set used in empirical modelling was used to determine the probability distribution of flyrock as given in Figure 5.4.

From Figure 5.4, it is established that the probability density function assumes the form of Weibull distribution and is given in Equation 5.6:

$$P(E) = 1 - e^{-\frac{R^n}{R_0}} \tag{5.6}$$

where R is the distance at which probability is to be worked, R_0 is the curve characteristic and n is the slope of the function.

The risk due to flyrock thus can thus be evaluated by using Equation 5.1, using the probability of flyrock defined by Equation 5.3 or Equation 5.5. The threat ratio T_r can be selected for a particular mining condition in conjunction with the probabilities to define the BDZ as shown in Figure 5.5 (includes the relative probabilities on different directions from the blast bench for flyrock). The BDZ thus defined (oblate in in Figure 5.5) accounts for risk zones only and can reduce from the event-based BDZ (circular in Figure 5.5).

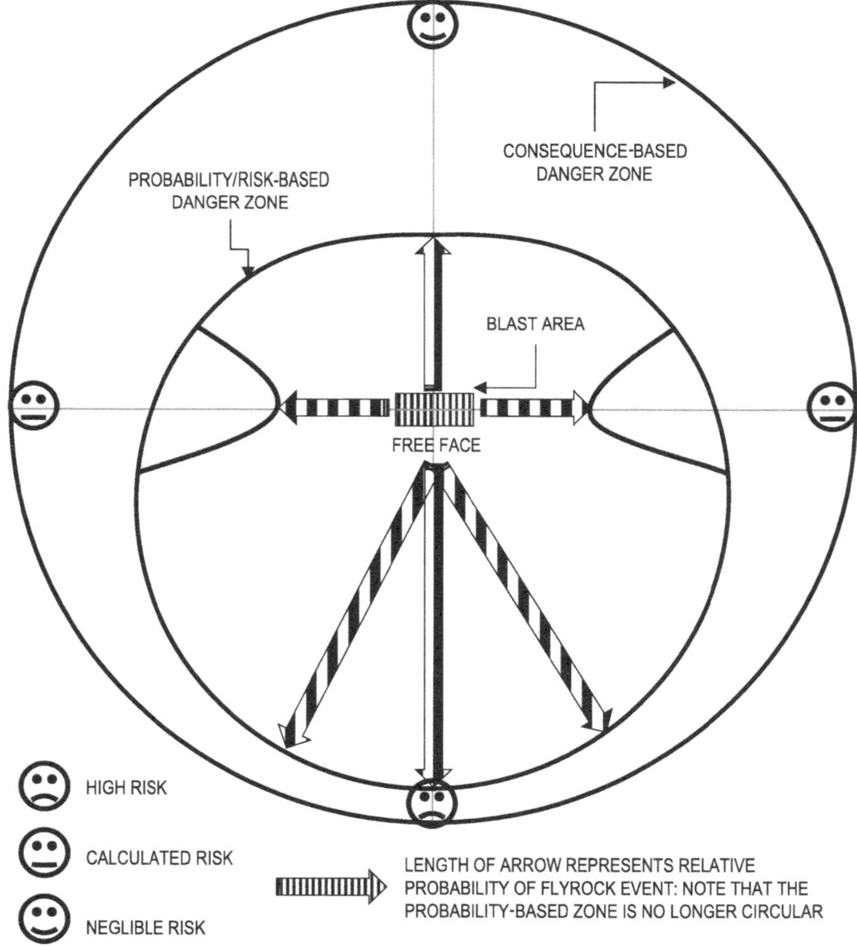

FIGURE 5.5 Risk-based blast danger zone in relation with the consequence-based zone.

This aforementioned method gives the blaster and the regulatory authorities a choice to alter permissible distance matching the geomining conditions and the associated probabilities of flyrock in a particular mine. This can be achieved through changes in blast pattern and monitoring of flyrock events and distance. If penalties or cost of an event is also known, the criterion translates into a scientific one to demarcate the blast danger zone and allows the BDZ to be different for different mines, and even difficult geomining conditions.

5.5.3 Pressure–Time Method

The pressure–time method (Section 4.7) can also be used for flyrock distance prediction due to its higher statistical significance. The method of St. George and Gibson (2001) discussed at Chapter 4, Equation 4.27, can be used for estimation of initial velocity of the flyrock and hence its distance. However, the model has several assumptions related to determination of the impact time, maximum distance at a launch angle of 45° questioned by McKenzie (2009), that need to be addressed before deploying the same. The use of approximate shape can be addressed by determining the shape of fragments achieved in breakage and flyrock. Once the range of flyrock is known, the BDZ can be defined with higher confidence by this method.

5.6 DEFINITION OF FLYROCK PERMISSIBLE DISTANCE

A proper definition permissible distance of flyrock should be based on properly evaluated blast design in conjunction with the OCs present. The risk-based method favours the general permissible distance of half (½) of the distance of OCs from the blast in the case of normal bench blasts. However, the same may not be applicable to places that are very sensitive in nature. Accordingly, the permissible distances can be worked out from the threat ratio (T_r) provided in Table 5.4 that can be used as a guide, where T_r is the ratio of flyrock permissible distance to the distance of OCs from the blast face.

Table 5.4 can be used as a guide to evaluate the risk provided the probabilities are established and penalty levels are known. The method can form a basis for future investigations.

TABLE 5.4
Definition of Flyrock Permissible Distance in Terms of Distance of OCs

S. No.	Nature of Working	Nature of Blasts	Nature of OCs	Flyrock Probability	Estimated Probability	Threat Ratio (T_r)	Risk Class Using T_r
1	Mines	Production, engineered	Habitats	Low to moderate		0.4–0.6	Calculated
			Equipment	Moderate		0.3–0.4	Minor
2	Construction site	Controlled	Habitats	Significant		0.2–0.3	No Risk
3	Sub-urban	Highly controlled	Habitats	High		0.1–0.2	No Risk

NOTES

1. https://central.bac-lac.gc.ca/.item?id=TC-BVAU-42948&op=pdf&app=Library&oclc_number=1032912260
2. www.ilo.org/wcmsp5/groups/public/---ed_dialogue/---sector/documents/normativeinstrument/wcms_617123.pdf
3. www.google.com/url?sa=t&rct=j&q=&esrc=s&source=web&cd=&ved=2ahUKEwjZroLjnb75AhU-RmwGHeYQCioQFnoECAgQAQ&url=https%3A%2F%2Fwww.gov.nl.ca%2Fmpa%2Ffiles%2FLGLUP-Avondale_DR.pdf&usg=AOvVaw2oJXQTTM2rQL4c7U4KkLsT
4. www.dgms.gov.in/writereaddata/UploadFile/Metalliferous%20Mines%20Regulation,%201961.pdf
5. www.dgms.gov.in/writereaddata/UploadFile/MMR%20Notification637186539489175908.pdf
6. https://environment.rajasthan.gov.in/content/dam/environment/RPCB/NGTORDERS/Original%20Application%20No.%2075_2020%20KHANAN%20GRASTH%20SANGHARSH%20SAMITI%20VS%20UNION%20OF%20INDIA.pdf
7. https://novascotia.ca/nse/dept/docs.policy/Guidelines-Pit-and-Quarry.pdf
8. www.law.cornell.edu/regulations/south-carolina/S-C-Code-Regs-89-150

REFERENCES

Adhikari, G. R. (1999). Studies on flyrock at limestone quarries. *Rock Mechanics and Rock Engineering*, *32*(4), 291–301. https://doi.org/10.1007/s006030050049

AEISG. (2011). *Code of good practice: Blast guarding in an open cut mining environment*. http://dmp.wa.gov.au/Documents/Dangerous-Goods/blast_guarding_cop_edition_1_march_2011.pdf

Arthur, J. D. (1979). Blasting safety requirements under the new surface mining law. *Mining Engineering*, 1568–1569.

Bajpayee, T. S., Bhatt, S. K., Rehak, T. R., Mowrey, G. L., & Ingram, D. K. (2003). Fatal accidents due to flyrock and lack of blast area security and working practices in mining. *Journal of Mines, Metals and Fuels*, *51*(11–12), 344–349.

Bajpayee, T. S., Rehak, T. R., Mowrey, G. L., & Ingram, D. K. (2002). A summary of fatal accidents due to flyrock and lack of blast area security in surface mining, 1989 to 1999. *Proceedings of the Annual Conference on Explosives and Blasting Technique*, *2*, 105–118.

Bajpayee, T. S., Rehak, T. R., Mowrey, G. L., & Ingram, D. K. (2004). Blasting injuries in surface mining with emphasis on flyrock and blast area security. *Journal of Safety Research*, *35*(1), 47–57. https://doi.org/10.1016/j.jsr.2003.07.003

Bandyopadhyay, P. K., Roy, S. C., & Sen, S. N. (2003). Risk assessment in open cast mining—an application of Yager's methodology for ordinal multiobjective decisions based on fuzzy sets. *Japan Journal of Industrial and Applied Mathematics*, *20*(3), 311–319. https://doi.org/10.1007/bf03167425

Blair, D. P. (2022). Probabilistic analysis of flyrock from blasting in surface mines and quarries. *International Journal of Rock Mechanics and Mining Sciences*, *159*, 105204. https://doi.org/10.1016/j.ijrmms.2022.105204

Blanchier, A. (2013). Quantification of the levels of risk of flyrock. In *Rock fragmentation by blasting, FRAGBLAST 10—proceedings of the 10th international symposium on rock fragmentation by blasting* (pp. 549–553). CRC Press/Balkema. https://doi.org/10.1201/b13759-77

Brnich, J. M. J., & Mallett, L. G. (2003). *Focus on prevention: Conducting a hazard risk assessment*. Pittsburgh Research Laboratory, National Institute for Occupational Safety and Health, 8p.

Carlson, D., & Eggerding, P. J. (2000). *Surface mine instructor reference and trainee review manual*. Michigan Technological University, 139p.

Davies, P. A. (1995). Risk-based approach to setting of flyrock "danger zones" for blast sites." *Transactions—Institution of Mining & Metallurgy, Section A*, (May–August), 96–100. https://doi.org/10.1016/0148-9062(95)99212-g

DGMS. (2003). *Technical circular*, DGMS(SOMA)/(Tech) Cir. No. 2 of 31st January 2003. Directorate General of Mines Safety, 2p.

Dick, R. A. (1979). A review of the federal surface coal mine blasting regulations. *International Society of Explosives Engineers*, 8p.

Kecojevic, V., & Radomsky, M. (2005). Flyrock phenomena and area security in blasting-related accidents. *Safety Science*, *43*(9), 739–750. https://doi.org/10.1016/j.ssci.2005.07.006

Little, T. N. (2007). Flyrock risk. *Australasian Institute of Mining and Metallurgy Publication Series*, 35–43.

Little, T. N., & Blair, D. P. (2010). Mechanistic Monte Carlo models for analysis of flyrock risk. In *Rock fragmentation by blasting—proceedings of the 9th international symposium on rock fragmentation by blasting, FRAGBLAST 9* (pp. 641–647). CRC Press/Balkema.

Loeb, J. T. (2012). *Regulatory mitigation of the adverse environmental effects of urban blasting*. University of British Columbia.

Lundborg, N. (1979). *The probability of flyrock damage* (SweDeFo Report No. DS1979:10). Swedish Detonic Research Foundation, 8p.

Lundborg, N., Persson, A., Ladegaard-Pedersen, A., & Holmberg, R. (1975). Keeping the lid on flyrock in open-pit blasting. *Engineering and Mining Journal*, *176*(5), 95–100. https://doi.org/10.1016/0148-9062(75)91215-2

McKenzie, C. K. (2009). Flyrock range and fragment size prediction. In *Proceedings of the 35th annual conference on explosives and blasting technique*, *2*, 17p. International Society of Explosive Engineers. http://docs.isee.org/%0AISEE/Support/Proceed/General/09GENV2/09v206g.pdf

Mine Health and Safety Act, 1996 (Act No. 29 of 1996) and Regulations, 587p (1998). www.hpcsa.co.za/Uploads/Legal/legislation/MINE_HEALTH_AND_SAFETY_ACT 29_OF_1996.pdf

Mishra, A. K., & Mallick, D. K. (2013). Analysis of blasting related accidents with emphasis on flyrock and its mitigation in surface mines. In *Rock fragmentation by blasting, FRAGBLAST 10—proceedings of the 10th international symposium on rock fragmentation by blasting* (pp. 555–561). CRC Press/Balkema. https://doi.org/10.1201/b13759-78

Nowrouzi, B., Rojkova, M., Casole, J., & Nowrouzi-Kia, B. (2017). A bibliometric review of the most cited literature related to mining injuries. *International Journal of Mining, Reclamation and Environment*, *31*(4), 276–285.

Raina, A. K. (2014). *Modelling the flyrock in opencast blasting under difficult geomining conditions*. Indian Institute of Technology—ISM. www.iitism.ac.in/pdfs/departments/mining/Research-degrees-completed.pdf

Raina, A. K., Chakraborty, A. K., Choudhury, P. B., & Sinha, A. (2011). Flyrock danger zone demarcation in opencast mines: A risk based approach. *Bulletin of Engineering Geology and the Environment*, *70*(1), 163–172. https://doi.org/10.1007/s10064-010-0298-7

Raina, A. K., Haldar, A., Chakraborty, A. K., Choudhury, P. B., Ramulu, M., & Bandyopadhyay, C. (2004). Human response to blast-induced vibration and air-overpressure: An Indian scenario. *Bulletin of Engineering Geology and the Environment*, *63*(3), 209–214. https://doi.org/10.1007/s10064-004-0228-7

Rehak, T. R., Bajpayee, T. S., Mowrey, G. L., & Ingram, D. K. (2001). Flyrock issues in blasting. *Proceedings of the Annual Conference on Explosives and Blasting Technique*, *1*, 165–175.

Richards, A., & Moore, A. (2004). Flyrock control—by chance or design. In *Proceedings of the 30th annual conference on explosives and blasting technique*, *1* (pp. 335–348). International Society of Explosive Engineers.

Roth, J. A. (1979). *A model for the determination of flyrock range as a function of shot conditions* (Report No. PB81222358). US Department of Commerce, NTIS, 61p.

Sari, M., Selcuk, A. S., Karpuz, C., & Duzgun, H. S. B. (2009). Stochastic modeling of accident risks associated with an underground coal mine in Turkey. *Safety Science*, *47*(1), 78–87. https://doi.org/10.1016/j.ssci.2007.12.004

Schwengler, B., Moncrieff, J., & Bellairs, P. (2007). Reduction of the blast exclusion zone at the black star open cut mine. *Australasian Institute of Mining and Metallurgy Publication Series*, 51–58.

St. George, J. D., & Gibson, M. F. L. (2001). Estimation of flyrock travel distances: A probabilistic approach. In *EXPLO 2001 conference* (pp. 409–415). AusIMM.

Verakis, H. C. (2011). Floyrock: A continuing blast safety threat. *Journal of Explosives Engineering*, 28(4), 32–37.

Verakis, H. C., & Lobb, T. E. (2003). An analysis of blasting accidents in mining operations. *Proceedings of the Annual Conference on Explosives and Blasting Technique*, *2*, 119–129.

Verakis, H. C., & Lobb, T. E. (2007). 2007G volume 1—flyrock revisited: An ever-present danger in mine blasting. In *33rd annual conference on explosives and blasting technique*, *1* (pp. 1–10). International Society of Explosive Engineers. http://docs.isee.org/ISEE/Support/Proceed/General/07GENV1/07v109g.pdf.

Zhou, Z., Xibing, L. I., Xiling, L. I. U., & Guoxiang, W. A. N. (2002). Safety evaluation of blasting flyrock risk with FTA method. *Safety*, *1*, 1184–1187.

6 Flyrock: "CAP IT"

Flyrock can be called as blaster's nightmare because it can damage, injure, and cause fatalties. Accordingly, there is a need to control the occurrence of the flyrock and at the same time restrict it to desired distances. To have a proper idea about flyrock control, it is essential to have knowledge of the cause of flyrock generation. Most of the mechanisms and reasons of flyrock occurrence have been outlined in Chapter 2. A comprehensive review of the reports of flyrock reveals that most of the reasons for flyrock occurrence have been listed by earlier works on the subject (Table 6.1). The causes of flyrock have been sorted in Table 6.1 to determine the maximum number of citations for each cause, as these amount to expert opinion. This is essential to enumerate the control measures of flyrock.

Among the causative factors the "geological anomalies" tops the list (Table 6.1) for generation of flyrock, followed by insufficient burden, insufficient stemming (length), inadequate delay, improper blast layout, excessive specific charge, poor quality of stemming, drilling accuracy, along with other causes. The occurrence of stemming length and its type are among the top causative factors of the flyrock causes. The observations, however, need to be put into perspective of variables used for prediction of flyrock distance (Table 6.2).

It may be pointed out that 51 research papers studied till date used artificial intelligence methods (see details of variables used in Table 4.4) for flyrock distance prediction. Since these publications do not introduce any new causes and use blast design variables only, for flyrock prediction, these have been grouped together in a single column under the heading "ANN, 2010–2022" in Table 6.2.

From Table 6.2, it emerges that stemming, burden, specific charge, spacing, hole depth, blasthole diameter, charge per delay, density of rock, specific drilling, linear charge concentration (density), and others appear to have been used for prediction of flyrock distance, most of the times, in order of their number of citations. It may be noted that maximum charge per delay does not feature in causes of flyrock as detailed in Table 6.1. A comparative analysis of the cause and prediction variables (Figure 6.1) indicates that geological anomalies have not been treated at all in flyrock distance (R_f) prediction models except by Raina et al. (2006). Also, the importance of variables presents a mismatch between the causative and predictive variables used for R_f prediction. Delay between the blastholes has been treated hardly, although there is mention of the same as a major causative variable. The reason for such anomalies is difficult to comprehend and demands introspection by the scientific fraternity.

Thus, defining the predictive regime of flyrock not only becomes somewhat ambiguous, but also difficult at the same time. However, it does not preclude exploring the preventive and control measures. Nonetheless, this analysis provides the researchers an opportunity to further investigate the problem, which may mean revisiting

DOI: 10.1201/9781003327653-6

TABLE 6.1
Causes of Flyrock as Reported by Various Authors in the Published Literature

Causes of Flyrock \ Reference	Schneider (1996)	Rehak et al. (2001)	Baliktsis and Baliktsis (2004)	Bhandari (1997)	Kuberan and Prasad (1992)	Mishra and Mallick (2013)	Zhou et al. (2002)	Mandal (1997)	Fletcher and D'Andrea (1987)	Davies (1995)	Opeyemi (2009)	Little (2007)	Verakis and Lobb (2007)	Kecojevic and Radomsky (2005)	Adhikari (1999)	Pradhan (1996)	Raina et al. (2011)	Amini et al. (2012)	Richards and Moore (2004)	Gupta (1990)	Fletcher and D'Andrea (1987)	Roth (1979)	Little and Blair (2011)	Bajpayee et al. (2004)	Shea and Clark (1998)	Ghasemi et al. (2012)	Workman and Calder (1994)	Kricak et al. (2012)	Total Citations
1. Geological anomalies	√	√	√	√	√	√		√	√	√	√	√	√	√			√	√			√	√	√	√	√	√	√	√	23
2. Insufficient burden	√	√	√	√	√	√	√	√	√	√		√	√	√	√		√	√	√	√	√	√	√	√					22
3. Insufficient stemming	√			√	√	√	√	√	√	√	√		√	√	√		√		√		√		√			√			17
4. Inadequate delay	√	√	√	√	√	√	√	√	√	√	√	√	√		√	√						√			√				17
5. Improper blast layout	√	√	√	√	√		√					√		√	√	√		√		√	√								15
6. Excessive specific charge	√	√	√		√		√		√	√	√		√	√		√	√										√		13

7. Poor quality of stemming	√	√	√													√			√	√		√			√				8
8. Drilling inaccuracy	√	√		√		√	√		√																				6
9. Poor confinement												√						√	√					√					4
10. Improper loading of explosives		√	√		√						√																		4
11. Misfire						√				√						√													3
12. Spacing								√							√														2
13. Density of explosive																				√									1
14. Waterlogged blasthole								√																					1
Total causes identified	9	8	7	7	7	6	6	6	6	6	5	5	5	5	5	5	4	4	4	4	4	4	3	3	3	2	2	1	

TABLE 6.2
Variables Used in the Prediction of Flyrock by Various Authors in Different Publications

Variables Used in Flyrock Distance Prediction \ Reference	ANN, 2010–2022 (Total 51) references)	Raina et al. (2011)	McKenzie (2009)	Stojadinović et al. (2011)	Richards and Moore (2004)	Roth (1979)	St. George and Gibson (2001)	Gupta (1990)	Lundborg (1974)	Total Citations
1. Stemming	46	1	1		1			1		50
2. Burden	46	1			1			1		49
3. Specific charge	43								1	44
4. Spacing	44									44
5. Hole depth	30	1								31
6. Charge per delay	26									26
7. Blasthole diameter	22	1	1						1	25
8. Density of rock	10			1		1	1			13
9. Linear charge concentration	9	1			1	1				12
10. Rockmass rating	8									8
11. Specific drilling	7									7
12. Velocity of detonation	3			1			1			5
13. Density of explosive	2		1	1			1			5
14. Drill/blasthole angle	3				1					4
15. Total charge length	1		1			1				3
16. Number of holes	3									3
17. Bench height	3									3
18. Subdrilling length	3									3
19. Uniaxial compressive strength	3									3
20. Hole-to-hole delay	3									3
21. Type of explosive	1					1				2
22. Fragment shape			1	1						2
23. Number of rows	2									2

Reference / Variables Used in Flyrock Distance Prediction	ANN, 2010–2022 (Total 51) references)	Raina et al. (2011)	McKenzie (2009)	Stojadinović et al. (2011)	Richards and Moore (2004)	Roth (1979)	St. George and Gibson (2001)	Gupta (1990)	Lundborg (1974)	Total Citations
24. Number of free faces	2									2
25. Volume of gaseous products	2									2
26. Geological strength index	2									2
27. Charge per blasthole	1									1
28. Rock conditions		1								1
29. Blasting index	1									1
30. Row-to-row delay	1									1
31. Volume of blasthole	1									1
32. Index of weatherability	1									1
Total variables used in predictions	4–12	6	5	4	4	4	3	2	2	

the whole phenomenon of flyrock and its different dimensions, from the inferences derived (Tables 6.1 and 6.2 and Figure 6.1).

Keeping in mind the aforementioned discussion, a proper layout of prevention of flyrock is attempted here, so that concerned persons make an organized effort to prevent the flyrock and achieve the objectives of their mine-mill fragmentation system (MMFS).

6.1 COMPLETE KNOW-HOW OF MINES

Two situations can be visualized in the case of mines. One, that a mine is yet to be opened, and the other that a mine is already operational and faces flyrock owing to presence of OCs. If a mine is yet to be opened, the management has an opportunity to decide upon several aspects of the mining, particularly the processes that are systemic in nature. For example, deciding on drill diameter, the shovel–dumper combination or other excavation machinery to be deployed in the mines, the orientation of the benches with respect to OCs, and even to reject drilling and blasting as a method of ore or overburden extraction should the situation warrant. The mines have option to design the face accordingly, to orient OCs towards less risk direction of the BDZ.

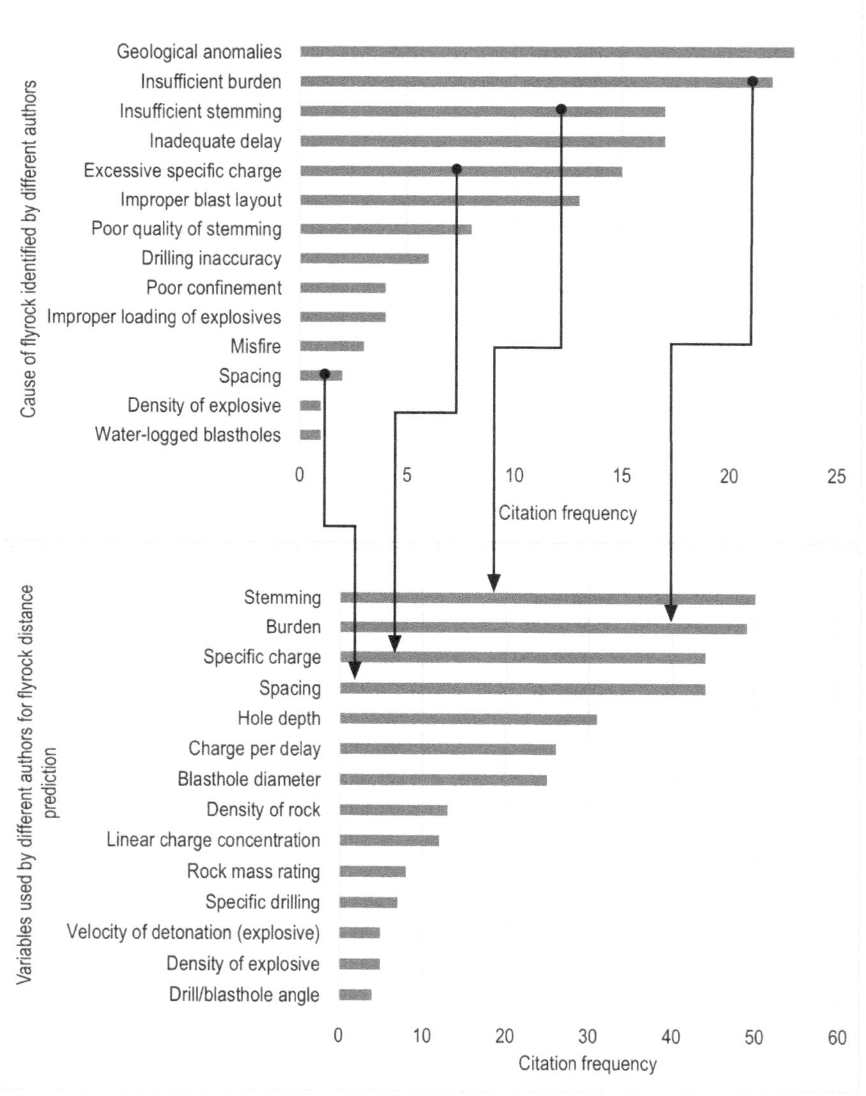

FIGURE 6.1 Comparison of causative and predictive variables of flyrock as investigated by different authors.

However, in both the cases mentioned earlier, it is essential to be aware of the mining method, the bench profiles, and the characteristics of the rockmass that is to be mined out. It must be remembered that there is a conflict in unit operations of the MMFS, as introduced in Chapter 1 (Figure 1.3), in relation with economics, modelling of fragmentation, throw, and flyrock. Few aspects of importance for flyrock prevention and control are described further.

6.1.1 Geology of the Mining Area

The geology of the mines is an important aspect of a blast design. An uncontrollable factor in blast design, it includes several variables that should be of interest to a blasting engineer. A detailed list of such variables along with their role in blasting are provided in Table 6.3.

TABLE 6.3
Geological Variables of Importance in Blast Design and Their Role in Blasting

S. No.	Geological Variables	Role in Blasting
1	Density of rockmass	Mechanics, breakage
2	Compressive strength (intact rock)	Breakage
3	Tensile strength (intact rock)	Breakage
4	In situ p-wave velocity (dynamic)	Rock response and breakage
5	Modulus of elasticity (dynamic)	Rock response and breakage
6	Poisson's ratio	Rock response and breakage
7	Number of joints	Defines the joint frequency, block size, throw, and influences flyrock distance
8	Joint spacing	Defines the block size related to the number of joints, fragmentation, influences flyrock distance
9	Joint orientation	Determines the shape of the blocks/flyrock, breakage
10	Joint condition	Defines the inter-block strength
11	Joint roughness	Defines the inter-block strength and even resistance to breakage
12	In situ block volume	Fragmentation, flyrock size
13	Water (sometimes controllable)	Influences the performance of explosives; influences stemming, and can induce flyrock in vertical direction
14	Unknown rock conditions (random); anomalies in rockmass	
	i. Presence of voids—specific to limestone	Flyrock, fragmentation
	ii. Presence of weak beds in competent formation	Flyrock
	iii. Presence of fissures in competent formation	Flyrock
	iv. Presence of in situ boulders in weak strata	
	v. Presence of breakage due to previous blasting	Flyrock

Most of the variables mentioned in Table 6.3 influence the blast performance in terms of fragmentation and throw. Therefore, it is necessary to have a complete know-how of the geological conditions, geo-mechanical characteristics of the rock-mass encountered in a mine to evolve the best blast design that conforms to the objectives of the MMFS, and yet prevent flyrock events.

6.1.2 Identification of Potential Zones of Flyrock

There are 14 conditions mentioned in Table 6.3 that may be responsible for flyrock. These represent the "geological anomalies" (Table 6.1) that surface at the top of the causative reasons of flyrock and are duly represented as the major cause of flyrock. However, a good deal of effort is required to log the geological anomalies in the rock-mass in the mines but is worth the exercise. These can be demarcated on the mining plan or blasting plan well in advance and verified at the time of blasting but before charging of blastholes. Any such anomaly will need to be addressed specifically during blasting. A geophysical survey can be of help not only to ensure detection of such rockmass defects, for total flyrock control, but also form a basis for devising proper blast design through impedance matching.

6.1.3 Presence of Objects of Concern

It is important to have a ready information about the houses, number of people residing nearby, if possible, their occupations and when they move out and move in from their houses, what livestock do they own and when the livestock can be in the near boundary of the mines, what types of assets do they own, and a broad information about the asset values. The relationship of residents in OCs and their psyche can be quite useful in deciding on blasting strategies.

6.1.4 Orientation of Objects of Concern with Respect to Blast Faces

The orientation of the OCs must be documented in a mine and their relationship with the blast face established in every blast. If the OCs are dynamic in nature, the possibility of these being towards front of free face of the blast should be evaluated and measures taken to clear the area or relocate these to other directions before conducting the blast.

In the case of static OCs, the probabilities of flyrock projected in that direction should be defined and evaluated before the blast. Risk management protocol should be specified in advance and measures to mitigate the risk taken, accordingly, before firing the blast.

6.1.5 Understanding the Cause-Effect Relationships

Once the baseline information as delineated in Sections 6.1.1–6.1.4 is available, it is prudent to have a brainstorming on the cause–effect relationships of flyrock. A logical analysis of citations presented in Table 6.1 can be of advantage to define the associations. The initial problem statement is "the occurrence of flyrock as is the only effect." The secondary effects will be impact of flyrock and its consequences. The causes can be thus related to the flyrock occurrence by individual mines by method of elimination and a protocol developed for its prevention and control. In this context, the fault tree analysis (Zhou et al., 2002) and flyrock probabilities introduced earlier and further expanded by Blanchier (2013) can be taken advantage of and used for analysing the phenomenon.

6.2 FLYROCK PREVENTION

Although prevention and control have similar connotation, a distinction is made over here by considering the former as a design aspect and the later as including all measures that can be taken to control the flyrock from hitting the OCs. This is the reason that both have been treated separately, despite significant overlap in reporting of such prevention and control measures in prevalent literature. Flyrock prevention can be possible with proper engineering where blasting engineer can devise a blast design that can be perfectly matched to mining conditions, geology, bench height, blasthole diameters, and prior information of the geological anomalies. The blast can be configured, and proper explosive selected to achieve the objectives defined by the management. It is important for the blaster to understand that every blast is a new blast and needs equal attention, so far as design and implementation is considered. The prerequisites for such design will be as mentioned in further sections.

A comprehensive flyrock prevention mechanism is difficult to compile. However, broad guidelines for flyrock prevention are provided in Chapter 7, Table 7.3. It is essential to understand that the preventive measures have a close relation with the blast design and understanding of the underlying causes (see Table 6.1) of flyrock. The details must include how and what is needed to prevent a flyrock occurrence.

6.2.1 Blast Face Survey

The first and foremost requirement for flyrock prevention is that a proper pre-blast face survey is conducted of the rock face that is intended for breakage by blasting. This will involve preliminary inspection and detailed mapping of the face for its geology and rock strength assessment, along with the joint spacing, length, and condition mapping. In addition, the presence of any natural anomalies must be recorded. Any damage to the blast face due to previous blast and its extent should be logged. Marking of blastholes and post-drilling measurements for burden, spacing, depth of the holes, and derivations for explosive loading of individual blastholes should be logged and communicated to blasting crew for strict implementation. Laser profiler surveys are of great value to the blaster that provides all measurements of B, S, l_{bh}, etc. that can be used to decide on the explosive quantity requirements of each blasthole. Modern drills are equipped with several sensors along with continuous data logging devices. Such drill data can be used to calculate the blasthole deviation for explosive control and to derive drilling index that can be correlated with the specific charge and required explosive energy calculations.

6.2.2 Proper Blast Design and Simulations

The pre-blast survey information obtained from the blast face can be used to engineer the blast, define the geometry, and fix the explosive configurations. Proper blast design means engineered definition of variables that are solely dependent on the blasthole diameter, rock type (blastability), explosive properties, and controlled by mine equipment specifications. A blasting engineer must bear in mind that the main reason for flyrock generation is the mismatch between the energy available and the work to be done (Little, 2007). Accordingly, the blast design must have a balance between the explosive energy available and work to be done. The procedure to make this happen has been discussed earlier in Section 4.8.3.

The stemming length, type, method, and burden require proper attention and should be as per design specifications. A complete description of the requirements of stemming and burden have been discussed earlier in Chapters 2 and 3.

Appreciation of the conflicting nature of unit operation costs of MMFS and optimization requirements is also necessary. For example, best results in blasting are obtained when the stiffness, i.e. H_b to B ratio, is more than 3 (Konya & Walter, 1991), which means that for a 110-mm diameter blasthole, the best bench height should be around 10–12 m with an explosive column length of 7.5–9 m. This means increase in specific charge that translates to higher charge per delay and charge per hole. The design will yield better fragmentation but higher vibrations, air overpressure, and greater flyrock distance. The bench height may also be restricted by loading equipment boom height. Since linear explosive distribution in blasthole is also important, the reduced bench height may mean reduction in design B and S to achieve good fragmentation but will increase the specific charge effecting the overall economics of the mine.

It is thus imperative that standard procedures for blast design and configuring it with a particular geomining condition through trials and simulations should be followed. Some basic guidelines for evolving such design have been provided in Chapter 2 and should be refined through additional inputs as provided in standard texts. Monitoring of blast outcome and matching the results with system objectives is necessary.

6.2.3 Air-decking: A Possible Prevention Case

Air-decking involves inserting an air gap between two charge decks in explosive column, hole bottom and explosive column, or explosive column and stemming material. Such decking is a special case of stemming (Rustan et al., 2011) and has been in practice for quite some time and benefits have been reported (Table 6.4). Air-decking thus distributes the explosive charge within the hole in special conditions that may require less explosive or firing of two decks in the explosive column at different time intervals. The method is believed to improve economics and breakage.

Whatsoever be the reason for the improvements and the underlying mechanism, one simple feature of the air-decking is that the charge per hole is reduced proportional to the length of the air-deck. This reduces the total explosive load in the blasthole. However, in the case of bottom or middle air-deck, the stemming length and related consequences will still apply. Even if the air-deck is applied and the

TABLE 6.4
Case Studies of Air-decking in Explosive Column Reported by Various Authors

S. No.	Author	Rock Type/ Description	d (m)	Deck Length	Hole Length	Deck Position	Results
1	Kabwe and Banda (2018)	Gneiss	0.165	1.8 (1.5 m air and 0.3 m gasbag)	12	Top	20% cost reduction in top column air-deck holes
2	Kabwe (2017)	Gneiss	0.165	1.8 (1.5 m airgap and 0.3 m gasbag)	12	Top	Improved fragmentation of 94% less than regular; top air-deck reduced explosives consumption significantly without loss in fragmentation or movement of the collar zone
3	Wu, Yu, Duan, et al. (2012)	Quartz sandstone and Muddy Shale	0.040	0.1	1.4	Top, middle, and bottom	Top air-deck charge induces tensile breakage in the zone. Middle air-decked charge is effective in cushion and pre-split. Lower air-deck helps in the expansion of the rock crack at the bottom of the blasthole
4	Wu, Yu, and Duan (2012)	Quartz sandstone and muddy shale	0.040	0.1	1.4	Top, middle, and bottom	Significant effect in breakage when charges are fired from top and bottom and middle of the explosive column.
5	Jhanwar (2011)	Review	–	–	–	Middle, top, bottom	Mid column air-deck preferred. Top air-deck for adequate breakage in the stemming regions, Bottom air-deck is generally not suggested except in the case of soft bottom regions
6	Fourney et al. (2006)	PMMA (Plexiglas) block	0.0127	0.254	0.005	Top and bottom	Pressure rise time is longer and improves performance

(Continued)

TABLE 6.4 (Continued)
Case Studies of Air-decking in Explosive Column Reported by Various Authors

S. No.	Author	Rock Type/ Description	*d* (m)	Deck Length	Hole Length	Deck Position	Results
7	Lu and Hustrulid (2003)	Granite	0.250	4.5	20	Top	Air-deck reduces the average final detonation pressure in the borehole and the degree of crushing near the borehole is reduced efficient explosive energy utilization
8	Jhanwar et al. (2000)	Manganese	0.1	0.9	8	Middle	Air decking in low strength moderately jointed rockmass gives good uniform fragmentation than conventional blasts
9	Jhanwar and Jethwa (2000)	Coal	0.15 and 0.25	1 + 1	10.2	Double decks in middle portion	Air-deck blasting technique is effective in soft and medium strength rocks
10	Jhanwar et al. (1999)	Manganese	0.105	0.9	7	Middle	Air-deck in the middle gives optimum fragmentation
11	Zarei et al. (2022)	Limestone and Schist	0.127	1–1.8	10–11.5	bottom	Substantial improvements in fragmentation and throw and ground vibrations

stemming is less than designed length, flyrock will still occur, irrespective of the deck. Hence, there is a need to evaluate the results of application of air-deck for flyrock control.

6.2.4 Misfires

Misfires are one of the common causes of flyrock, particularly if a delay fails in the front row and the following rows fire. When blasting with non-electric shock tube (NeSt) delay combination, it is always recommended to connect blastholes in the first row to subsequent rows using "row-to-row" connections, rather than connecting all the holes in a row and separating rows with a single delay. This method of connection ensures that progression of detonation of other holes in sequence is stopped, if a blastholes fails to initiate, hence preventing flyrock. However, it is important to know the reliability of detonators beforehand by seeking such reports from the manufacturer during procurement and conducting random tests on such detonators with precision measurement methods.

6.2.5 Risk-Based Prevention

The probability of flyrock distance exceeding the permissible limit or BDZ can be estimated with using the probability density function or probability of failure of a safety rule as explained in Chapter 5. A threat ratio, T_r, a ratio of the permissible distance of flyrock and the distance of OCs, can be used for risk due to flyrock calculation, which is a product of the probability of a flyrock event and the threat ratio as defined earlier. The process example is provided in Figure 6.2, wherein five risk

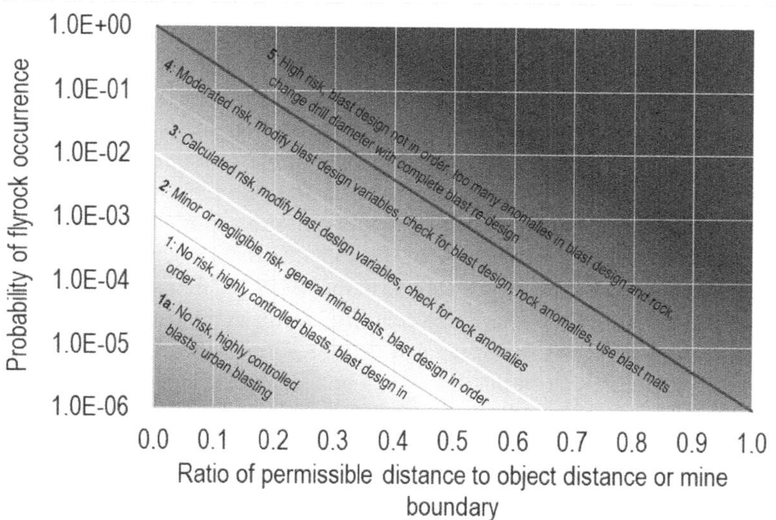

FIGURE 6.2 Plot of probability of a flyrock event versus threat ratio for risk classification and suggested blasting method for different risk classes.

classes have been defined ranging from "No risk" to "High risk." Based on the category of risk thus defined, the blast design variations or systemic changes to be exercised for flyrock prevention have been defined in Figure 6.2.

Figure 6.2 helps to identify different risk classes namely A to E that are defined as follows:

1a: *No risk*: This class indicates that blasting practice is highly controlled. Such practices can be followed while blasting very near to the structures, particularly, in urban areas.

1: *Negligible Risk:* The class defines negligible or acceptable risk and represents bench blasting in difficult geomining conditions. Explosive variations can be used to shift risk to Class 1a, if required.

2: *Calculated Risk*: This class represents calculated risk where trade-off between flyrock risk and productivity can be employed. To shift the class of risk to classes 1a or 1, controlled measures like modifications in blast design and explosive variations can be attempted.

3: *Moderate Risk:* This class represents the situations where probability of flyrock is moderate and definite changes in blast practice/design are required. Also, if blasting is to be continued as practiced, then blast mats will be essential to cap the flyrock.

4: *High Risk*: This risk class points to a total shift in blasting practice and even a systemic change in drill diameter and equipment configurations will be essential.

6.3 FLYROCK CONTROL

Flyrock control as appears in the public domain (Table 6.5) falls in different categories, viz. blast design, face survey, defining the BDZ, communication, and practice. Out of 20, 2–9 measures have been recommended by domain experts to control the flyrock with a significant overlap with preventive method.

Nine such measures in Table 6.5 are not related to blast design but represent training, communication, management, and implementation or blasting practice. Some of the observations are group variables, some are subcomponents or independent variables, and others represent behavioural or operational aspects. One of the issues with such reporting, in most of the cases, is inclusion of area security as cause or preventive measure. Neither does "area security" result in a flyrock, nor does it help to prevent the flyrock event. Blast area security is simply a measure to mitigate the risk due to flyrock and can be included as management or legislative measure and not a design criterion and can thus be regarded as a control method. Accordingly, the flyrock control measures have been further classified into different groups to have a clarity on the subject and at the same time evaluate these in terms of measures to evaluate each component along with their relationship with flyrock. These measures are compiled and presented in Table 7.4. The process is intended to help in defining the control measures of flyrock. A thorough examination of Table 6.5 can be useful in defining the flyrock control methods.

TABLE 6.5
Flyrock Control Measures of Flyrock Suggested by Various Authors

Reference / Preventive measures	Proper Stemming/material	Proper blast design	Area security	Knowledge of geology	Drilling accuracy	Proper burden	Communication	Training and education	Preloading inspection	Explosive confinement	Blasthole Loading Method	Pre-drilling Inspection	Muffling	Hazard Assessment	Bench Height	Proper Initiation Method	Bore Hole Diameter	Delay Timing	Post Firing Inspection	VoD of Explosive	Total Methods Identified
	1	2	3	4	5	6	7	8	9	10	11	12	13	14	15	16	17	18	19	20	
Thote et al. (2013)	√	√	√	√	√					√											6
Mishra and Mallick (2013)			√																		1
Kricak et al. (2012)	√		√	√		√	√			√		√		√							8
Verakis (2011)	√		√	√		√	√	√	√					√				√			9
Opeyemi (2009)	√	√	√	√			√					√							√		7
Rathore and Jain (2007)	√	√			√																3
Verakis and Lobb (2007)	√	√	√	√			√		√			√									7
Kecojevic and Radomsky (2005)			√					√													2
Baliktsis and Baliktsis (2004)	√	√	√	√		√	√	√							√						8
Zijun (2002)		√	√	√			√	√	√			√		√							8
Rehak et al. (2001)		√		√	√						√										4
Schneider (1996)		√	√		√			√						√							5
Mandal (1997)		√	√		√	√		√					√		√	√	√				9

(Continued)

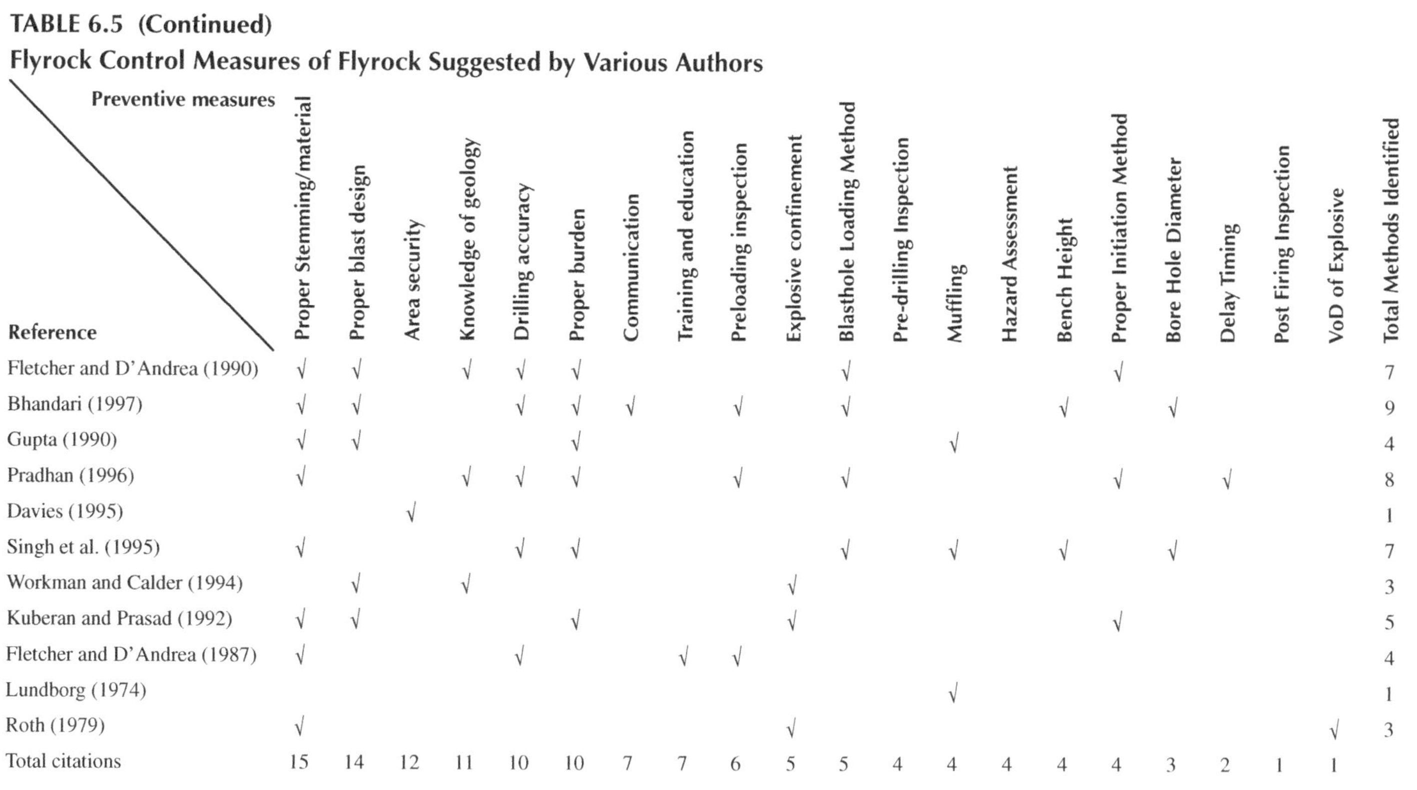

TABLE 6.5 (Continued)
Flyrock Control Measures of Flyrock Suggested by Various Authors

Reference \ Preventive measures	Proper Stemming/material	Proper blast design	Area security	Knowledge of geology	Drilling accuracy	Proper burden	Communication	Training and education	Preloading inspection	Explosive confinement	Blasthole Loading Method	Pre-drilling Inspection	Muffling	Hazard Assessment	Bench Height	Proper Initiation Method	Bore Hole Diameter	Delay Timing	Post Firing Inspection	VoD of Explosive	Total Methods Identified
Fletcher and D'Andrea (1990)	√	√		√	√	√					√					√					7
Bhandari (1997)	√	√			√	√	√		√		√				√		√				9
Gupta (1990)	√	√				√							√								4
Pradhan (1996)	√			√	√	√			√		√					√		√			8
Davies (1995)			√																		1
Singh et al. (1995)	√				√	√					√		√		√		√				7
Workman and Calder (1994)		√		√						√											3
Kuberan and Prasad (1992)	√	√				√				√						√					5
Fletcher and D'Andrea (1987)	√				√			√	√												4
Lundborg (1974)													√								1
Roth (1979)	√									√										√	3
Total citations	15	14	12	11	10	10	7	7	6	5	5	4	4	4	4	4	3	2	1	1	

6.3.1 Area Security Issues

Area security deals with the BDZ and guarding the area demarcated as danger zone around the blast face. It is necessary to have a proper protocol for ensuring that no movable assets are within the BDZ, no livestock or humans are present in the zone at the time of blasting. The level of security can even be determined with the help of the risk classes, as defined in Section 6.2.5. Proper guarding of the BDZ is essential and the officer authorized to order firing of blasts should seek all clearances from people guarding the area. Other requirements have been discussed earlier at relevant sections.

D'Andrea and Fletcher (1983), while examining 369 case studies of mining accidents, reported that most accidents occurred during scheduled blasting, and failure of the blast area security system was the major cause. Failure to clear the blast zone, failure of personnel to follow the protocol, insufficient guarding, and failure to retreat to a safe location or take adequate cover were the causes related to BDZ. Other common reasons for accidents were premature blasts, excessive flyrock, fumes, and misfires. The blaster in most of the accidents was injured by such events.

6.3.2 Risk Criterion and Risk Management

Risk in engineering is a difficult subject as it involved complicated relationships within cause, effect, and consequences of an event that results in accidents, injuries, and fatalities. Risk is a measure of loss due to human activity and its analysis process involves characterizing, managing, informing, magnitude of risk along with determination of contributing factors, and uncertainties of loss (Modarres, 2006). The three main constituents of risk analysis are risk assessment, risk management, and risk communication. In the case of flyrock, the evaluation of risk and its application has been detailed earlier. The management of risk due to flyrock will hence be covered over here. The communication part will be described later.

Few case studies that pertain to risk management in the domain of blasting include new method of risk management (Raina & Bhatawdekar, 2022), qualitative and quantitative risk evaluation and management method (Little, 2007), risk management near concrete structures (Nieble & Penteado, 2016), risk assessment methodology for blasting (Baliktsis & Baliktsis, 2005; Seccatore et al., 2013), flyrock prevention through risk management (Baliktsis & Baliktsis, 2004), and application of artificial intelligence techniques in risk analysis and management (Bhatawdekar et al., 2021).

Flyrock risk management will principally involve strategies or activities to control/prevent flyrock with a major focus to prevent any accidents. Comprehensive detail of the risk management for explosive-related operations involves proactive and reactive approaches to risk management (Diacolabrianos & Twining, 2007) that correspond to preventive and control measures in the case of flyrock. A modern approach to flyrock risk management has been given by Raina and Bhatawdekar (2022). A risk management approach for blasting is provided in Figure 6.3 that applies to blasting and related outcomes as flyrock cannot be seen in isolation. However, as demonstrated earlier, the risk will need to be identified while devising risk management policies.

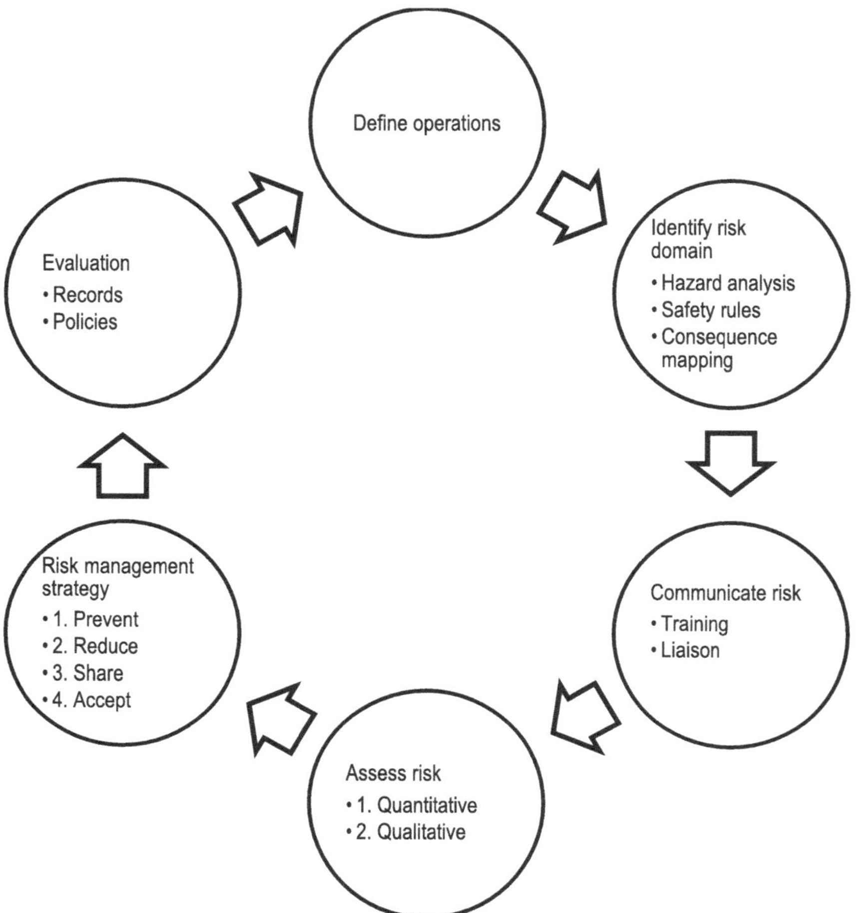

FIGURE 6.3 Risk management process in blasting.

Once the risk due to flyrock is calculated in relative or absolute scale, then the management has the responsibility to define a policy either to avoid, control, accept, or mitigate the risk. The decision in our case must be made in line with legal requirements. These policies may correspond to the risk classes defined in risk criterion for flyrock earlier.

When decisions and policies are in place in a mine, the management must define a proper protocol to achieve the objectives of the policy. Being aware of the legalities, the best practice will be to (1) continuously assess the risk and risk management practice, (2) redefine the risk management strategies, if the risk is high, (3) continuously assess the effectiveness of risk management strategies with changes in mining conditions and progression of mining activities towards the OCs.

It is essential to understand that a human factor is always in place that results in accidents. Brulia (1993), while referring to study on accidents (D'Andrea & Fletcher, 1983), mentioned that about 80–90% of all accidents are caused by human factors while listing five major reasons that contributed to these accidents that have been rearranged and explained here:

1. *Lack of Planning:* The knowledge of a blasting operation is essential. In absence of the comprehension, mistakes are obvious.
2. *Inadequate Instruction:* Lack of proper communication of proper methodology and deployment of untrained or improperly trained personnel in blasting will produce poor results and accidents.
3. *Negligence:* Any slack or failure to observe safety rules, non- or miscommunications about safety policy and adherence to rules can result in accidents.
4. *Hasty Decisions:* Proper thought and introspection of a blasting situation is required by the blasting engineer. Haste in implementation of instructions without proper thought amounts to taking hazardous shortcuts.
5. *Overconfidence:* Ignorance of risk due to blasting can be one of the causes of flyrock and blasting-related accidents as it amounts to taking unaccounted risk/chances.

All the aforementioned factors lead to increased risk of accidents and thus should be given proper weightage and attention by the planners and management of a mine to have a proper control over the blasting operations and safety.

6.4 DO ELECTRONIC DETONATORS PRESENT A CASE?

Electronic detonators (EDs) have been in vogue for quite some time with their benefits (Cardu et al., 2013). The basic mechanics of improvements in blasting with the use of electronic detonators has been investigated by several authors, e.g. Johnson (2014), Rossmanith (2002), Sjöberg et al. (2012), and Yi et al. (2016). Some of the advantages of such detonators have been documented by Mangum and Bryan (1987). There are sufficient evidences that electronic detonators can be used to control the fragmentation and hence optimize the mine-mill fragmentation system (Bilodeau et al., 2008; Chavez et al., 2007; Kanchibotla & Watson, 2004). The best and logical benefit of such detonators is the control of ground vibration and air overpressure, as reported by several authors, e.g. Mishra et al. (2017), Sharma and Rai (2016), and Singh et al. (2019, 2021). Even the psychological impact on the people with the use of electronic detonators is documented (Raina et al., 2015).

However, the documentation of flyrock with electronic detonators presents a cursory mention of flyrock control. The manufacturers claim that such detonators provide total flyrock control but the scientific documentation and mechanism of control of flyrock is lacking. The reasons for the control of flyrock with EDs could be related to proper delay application with millisecond control, its reliability, true bottom hole initiation, and minimal time scatter in delays. The cut-offs are also eliminated, misfires can be minimal, and initiation sequence is followed as planned with EDs.

This presents the advantage of EDs over NeSt where wrong connections can result in flyrock, and reliability of timing is impaired due to the in-hole long delay and shock tube lengths.

In several blasts, monitored by the author, the flyrock could be controlled using precise delays using electronic detonators, the incidences of flyrock were still there. Hence, the case for electronic detonators as a total flyrock control needs validation.

6.5 TOTAL FLYROCK CONTROL WITH BLAST MAT

Muffling of blast faces with different methods is a common practice in controlled excavations. Many a times, the blast face, particularly the top surface, is covered with used conveyor belts and sandbags (Figure 6.4a and b), used tyres, steel or polyethylene cable nets (Figure 6.4b), or blast mats (Figure 6.4c) or creating barricades (Figure 6.4d and e), or used dumper tyres on top of the blasthole (Figure 6.4f) to arrest the flyrock at its origin.

FIGURE 6.4 Types of muffling/blast mats used in blasting.

The function of a blast mat is to prevent flyrock from leaving the blast face. Rustan et al. (2011) provided a mantra for total flyrock control by blast mat:

"The best way to cap the flyrock is to use a proper blast mat."

The term "proper" used in the aforementioned definition is qualitative. However, a blast mat must conform to the following requirements:

1. It should be strong and compact enough to arrest any flyrock and prevent it from leaving the blast face.
2. During the blasting, the blast mat should not be thrown away from the blast face, which means that it should have sufficient weight.
3. It should be economical and should sustain wear and tear with good life span.

There are practically negligible published cases on standards and requirements of a blast mat. The evaluation of blast mat has been generally qualitative in nature and hence standard test was developed by Coy and Worsey (2015a) while calculating forces witnessed by blast mats through crater formation, initial velocity of flyrock, and fragmentation. Coy and Worsey (2015b), simultaneously, devised and tested a blast mat devised from tractor rubber treads. A fundamental study by Lovtsov and Kostiunina (2017) on gas permeable blasting mats, demonstrated through numerical and full-scale experiments that the axial forces in elastic elements of the blast mat and its maximum flight height was a function of the detonation sequence of the blastholes. There are lot of examples where authors, viz. Schneider (1996, 1997), Rathore and Jain (2007), and Venkatesh et al. (2013), demonstrated that flyrock can be prevented by the use of blast mats.

Blast mats are available on the market, and most of these are made from rubber or used tyres, except for some specialized ones that are made from polymer or steel wires. There are examples of woven wire mats that the manufacturers claim to have better resistance to blast forces and easily vent the blast gases. The general specifications of the blast mats are mentioned in Table 6.6.

There are few main concerns in use of blast mats:

1. These are quite heavy and difficult to handle. Proper equipment is required to place the mat on the blast face. Several mats are required depending on the area of the blast.
2. The blast mat is laid on charged and connected holes. Movement of a heavy equipment required for placement of blast mats over or near the blastholes can be unsafe.
3. The blast mat placement and quantity required does not favour its use in production (mining) blasts as it places constraints on the resources, including manpower, time, and cost.

Nonetheless, the use of blast mat is a good choice for blasting in urban areas or in benches where flyrock cannot be controlled by engineered blast design modifications and the flyrock risk is too high.

TABLE 6.6
Types, Material Used, Dimensions and Weight of Blast Mats for Flyrock Control

Mat Brand	Mat Name (as Appears in Websites)	Material	Dimension (m)	Weight (kg)
A	Control blasting mats	Sliced-up rubber tyre scrap, bound together with ropes, cables, and clamps	2.74 × 3.05 × 0.25	1150–1200 kg approximate
B	Tyre blasting mats	Used tyres	2.74 × 3.05 (thickness not known)	Not provided
C	Blasting mats	Sliced-up rubber tire scrap bound together with ropes, cables, and clamps	2.74 × 3.05 (thickness not known)	1200
D	Tyre blasting mats	Rubber tyre	2.74 m × 3.66 m × 0.20	Up to 1700
E	Heavy duty black blasting mats	Rubber tyre	2.74 × 3.05 × 0.08	Up to 1000
F	Blasting mats	Rubber	3.05 × 3.68 (thickness not known)	1700
G	Tyre blasting mats	Rubber tyre	Thickness 0.08 (other dimensions not known)	1000

6.5.1 Case Study

A case study of flyrock control by Raina et al. (2016) using blast muffling and mats for excavating an irrigation canal in basaltic formation is presented over here. Mud houses were situated at 18–30 m, and an old overhead tank was located at 13 m from the excavation. The houses and overhead tank both were to be protected from ground vibrations, air overpressure, and flyrock. In tune with the requirements, few trials were conducted away from the structures and later a perfect design was formulated for controlled blasting. Relevant details of the study are provided in Table 6.7. Ground vibration and air overpressure predictions and hence charge per delay calculations were made with the help of site-specific attenuation relationships and flyrock was capped with the help of multi-tier muffling arrangement. The blast surface was first covered with sandbags to prevent damage to the surface firing connections (shock tubes), that in turn provided additional cover. The blast mats were placed over the sandbags. A BDZ of 300 m was decided for residents, who were moved beyond this zone before firing the blast. A few images of the structures, muffling practice, and actual blasts are shown in Figure 6.5.

The aforementioned case study is an excellent example of total flyrock control by use of muffling arrangements.

TABLE 6.7
Details of Working, Blast Design, and Explosive Configuration Used in Blasting Near Houses

Blast Design Variables	Unit	Value	Distance of Structure from Blast (m)	Maximum Charge Per Delay/Hole (kg)
Blasthole diameter	m	0.075	18	1.3
Explosive diameter	m	0.05	20	1.6
Drill depth	m	3–5 (max)	22	2
Burden	m	2	24	2.4
Spacing	m	2.5	26	2.8
Stemming length	m	2.3 (l_s/B = 1.15)	30	3.7

FIGURE 6.5 Some images of the case study with blast mats in a very near structure blasting. (a) Image of a mud house near the excavation. (b) View of the overhead tank near the blast area. (c) Method of placing blast mats on the blast top surface. (d) Placement of muffling and blast mat over the blast surface in progress. (e and f) View of blasting in progress (no flyrock is seen to eject from the top or front of the blast).

6.6 REPORTING

A need exists to change the existing rules on flyrock reporting. Flyrock(s) that travel more than twice of the bench width of the mine should be reportable and should not invoke penalties. A complete description of such events should be logged in proper blasting database consisting of rockmass and blast design details and reason of flyrock, along with the size, dimensions, and flyrock distance. A video of the blast taken along the bench strike from a safe distance is recommended. This will ensure proper reporting and hence proper documentation of the flyrock distances along with other important blasting information. A proper documentation of flyrock events and video analysis details will help the miners and researchers to focus on the probabilities of flyrock and accordingly devise a risk analysis scheme for a mine. An international collaboration through an established organization can help to document and define flyrock probabilities and put causes and preventive measures into perspective and subject these to rigorous analysis.

6.7 COMMUNICATION AND TRAINING

Communication is primarily the domain of safety professionals who often use a multifaceted approach for injury prevention. In the case of blasting, the approach is as follows:

1. Definition of site-specific policies and procedures to implement the same. This will involve communication of the blast design for every blast well in advance to the drillers and blasters. This will also involve communication of the risk associated with blasting in general and specific to a particular mine.
2. Continuous training and education of personnel involved in drilling and blasting, which may even be necessitated by the change in persons responsible for unit operations like drilling and blasting. A need exists for educating drillers for drilling in accordance with the approved blast pattern and blasters for following the survey and charging of blastholes as per the approved design. Communication of scientifically validated knowledge with respect to blasting in general and flyrock should be part of the training. Information on the subject can be traced to Sen and Hayward (2007).
3. Proper buy-in and commitment of drilling and blasting personnel. This will mean evaluating their performance in accordance with the degree of implementation of the safety policies and other guidelines, in addition to general rules and regulations.

A prior intimation to all stakeholders in blasting and people living nearby should be a practice. Generally, the time of blasting is fixed in all mines. However, any changes in the timing should be intimated to all such persons. This may mean establishment of prior liaison with the people living about the mines by the human resource and safety department of the mines. Building a cordial relationship with the people living in adjacent localities should be a priority to mitigate the human response and reactions.

Proper guarding also is an important component of safety from flyrock. The blast danger zone should be clearly marked and communicated to the people and persons responsible in the mines should be thorough with the procedures, thereof. During and after blast, alerts through sirens and proper communication between people responsible for securing the area is a compulsory step. The guards should be thoroughly educated and in knowledge of dangers of flyrock.

REFERENCES

Adhikari, G. R. (1999). Studies on flyrock at limestone quarries. *Rock Mechanics and Rock Engineering*, *32*(4), 291–301. https://doi.org/10.1007/s006030050049

Amini, H., Gholami, R., Monjezi, M., Torabi, S. R., & Zadhesh, J. (2012). Evaluation of flyrock phenomenon due to blasting operation by support vector machine. *Neural Computing and Applications*, *21*(8), 2077–2085. https://doi.org/10.1007/s00521-011-0631-5

Bajpayee, T. S., Rehak, T. R., Mowrey, G. L., & Ingram, D. K. (2004). Blasting injuries in surface mining with emphasis on flyrock and blast area security. *Journal of Safety Research*, *35*(1), 47–57. https://doi.org/10.1016/j.jsr.2003.07.003

Baliktsis, E., & Baliktsis, A. (2004). Flyrock risk prevention—from theory and ideas to a perfectly applied blasting project. In Z. Agioutantis & K. Komnitsas (Eds.), *Proceedings of the 1st international conference on advances in mineral resources management and environmental geotechnology* (pp. 17–23). Heliotopos Conferences.

Baliktsis, E., & Baliktsis, A. (2005). Aiming to the perfect blasting result: Reality or just a Utopia? Facts, parameters and thoughts based on two case histories from EGNATIA ODOS—Ring Road of Salonica section, in Northern Hellas. In *5th international scientific conference of modern management of mine producing, geology and environmental protection, SGEM 2005* (pp. 199–209). International Multidisciplinary Scientific GeoConferences.

Bhandari, S. (1997). *Engineering rock blasting operations*. Balkema.

Bhatawdekar, R. M., Armaghani, D. J., & Azizi, A. (2021). Applications of AI and ML techniques to predict backbreak and flyrock distance resulting from blasting. In *Environmental issues of blasting: SpringerBriefs in applied sciences and technology* (pp. 41–59). Springer. https://doi.org/10.1007/978-981-16-8237-7_3

Bilodeau, M., Labrie, D., Boisclair, M., Beaudoin, R., Roy, D., & Caron, G. (2008). Impact of electronic blasting detonators on downstream operations of a quarry. *Minerals and Metallurgical Processing*, *25*(1), 32–40. https://doi.org/10.1007/bf03403383

Blanchier, A. (2013). Quantification of the levels of risk of flyrock. In *Rock fragmentation by blasting, FRAGBLAST 10—proceedings of the 10th international symposium on rock fragmentation by blasting* (pp. 549–553). CRC Press/Balkema. https://doi.org/10.1201/b13759-77

Brulia, J. C. (1993). Quality: A new approach to improve blasting safety. In *Proceedings of the annual conference on explosives and blasting technique* (pp. 55–64). International Society of Explosives Engineers.

Cardu, M., Giraudi, A., & Oreste, P. (2013). Uma revisão das vantagens dos detonadores eletrônicos. *Revista Escola de Minas*, *66*(3), 375–382. https://doi.org/10.1590/S0370-44672013000300016

Chavez, R., Leclercq, F., & Jurg, A. (2007). Modern blasting using electronic detonators and the total drill and blast concept in quarries. *Australasian Institute of Mining and Metallurgy Publication Series*, 121–127.

Coy, M. K., & Worsey, P. N. (2015a). *Development of a blasting mat test* (p. 12p). International Society of Explosives Engineers.

Coy, M. K., & Worsey, P. N. (2015b). Testing of a new blasting mat constructed using rubber tractor treads. *International Society of Explosives Engineers*, 1–10.

D'Andrea, D. V., & Fletcher, L. R. (1983). Analysis of recent mine blasting accidents. *International Society of Explosives Engineers*, 105–122.

Davies, P. A. (1995). Risk-based approach to setting of flyrock "danger zones" for blast sites." *Transactions—Institution of Mining & Metallurgy, Section A*, (May–August), 96–100. https://doi.org/10.1016/0148-9062(95)99212-g

Diacolabrianos, E., & Twining, D. J. (2007). A control measures based approach to risk management. In P. Moser (Ed.), *Vienna conference proceedings* (pp. 377–385). European Federation of Explosives Engineers. https://doi.org/ISBN 978-0-9550290-1-1

Fletcher, L. R., & D'Andrea, D. V. (1987). *Reducing accident through improved blasting safety*. USBM IC, 9135 (pp. 6–18). United States Bureau of Mines.

Fletcher, L. R., & D'Andrea, D. V. (1990). Control of flyrock in blasting. *Journal of Explosives Engineering*, *7*(6), 167–177.

Fourney, W. L., Bihr, S., & Leiste, U. (2006). Borehole pressures in an air decked situation. *Fragblast*, *10*(1–2), 47–60. https://doi.org/10.1080/13855140600858198

Ghasemi, E., Sari, M., & Ataei, M. (2012). Development of an empirical model for predicting the effects of controllable blasting parameters on flyrock distance in surface mines. *International Journal of Rock Mechanics and Mining Sciences*, *52*, 163–170. https://doi.org/10.1016/j.ijrmms.2012.03.011

Gupta, R. N. (1990). Surface blasting and its impact on environment. In R. K. Trivedy & M. P. Sinha (Eds.), *Impact of mining on environment* (pp. 23–24). Ashish Publishing House.

Jhanwar, J. C. (2011). Theory and practice of air-deck blasting in mines and surface excavations: A review. *Geotechnical and Geological Engineering*, *29*(5), 651–663. https://doi.org/10.1007/s10706-011-9425-x

Jhanwar, J. C., Cakraborty, A. K., Anireddy, H. R., & Jethwa, J. L. (1999). Application of air decks in production blasting to improve fragmentation and economics of an open pit mine. *Geotechnical and Geological Engineering*, *17*(1), 37–57. https://doi.org/10.1023/A:1008899928839

Jhanwar, J. C., & Jethwa, J. L. (2000). The use of air decks in production blasting in an open pit coal mine. *Geotechnical and Geological Engineering*, *18*, 269–287. https://doi.org/10.1023/A:1016634231801

Jhanwar, J. C., Jethwa, J. L., & Reddy, A. H. (2000). Influence of air-deck blasting on fragmentation in jointed rocks in an open-pit manganese mine. *Engineering Geology*, *57*(1–2), 13–29. https://doi.org/10.1016/S0013-7952(99)00125-8

Johnson, C. E. (2014). *Fragmentation analysis in the dynamic stress wave collision regions in bench blasting*. Ph.D. Thesis, University of Kentucky, 158p.

Kabwe, E. (2017). Improving collar zone fragmentation by top air-deck blasting technique. *Geotechnical and Geological Engineering*, *35*(1). https://doi.org/10.1007/s10706-016-0094-7

Kabwe, E., & Banda, W. (2018). Stemming zone fragmentation analysis of optimized blasting with top-column air decks. *CIM Journal*, *9*(1), 51–63. https://doi.org/10.15834/2018.1

Kanchibotla, S. S., & Watson, J. (2004). Electronic detonators—following a scientific process to sustain maximum value. *Australasian Institute of Mining and Metallurgy Publication Series*, 139–145.

Kecojevic, V., & Radomsky, M. (2005). Flyrock phenomena and area security in blasting-related accidents. *Safety Science*, *43*(9), 739–750. https://doi.org/10.1016/j.ssci.2005.07.006

Konya, C. J., & Walter, E. J. (1991). *Rock blasting and overbreak control* (No. FHWA-HI-92–001; NHI-13211). National Highway Institute, 430p.

Kricak, L., Kecojevic, V., Negovanovic, M., Jankovic, I., & Zekovic, D. (2012). Environmental and safety accidents related to blasting operation. *American Journal of Environmental Sciences*, *8*(4), 360–365. https://doi.org/10.3844/ajessp.2012.360.365

Kuberan, R., & Prasad, K. K. (1992). Environmental effects of blasting and their control. *Proceedings of the Workshop on Blasting Technology for Civil Engineering Projects*, *2*, 145–159.

Little, T. N. (2007). Flyrock risk. *Australasian Institute of Mining and Metallurgy Publication Series*, 35–43.

Little, T. N., & Blair, D. P. (2011). The influence of geology on blasting hazards and results. In *Proceedings of the 8th international mining geology conference 2011* (pp. 341–352). AusIMM.

Lovtsov, A. D., & Kostiunina, O. A. (2017). Dynamic analysis of gas permeable blasting mat as geometrically nonlinear system with unilateral constraints. *Procedia Structural Integrity*, *6*, 122–127. https://doi.org/10.1016/j.prostr.2017.11.019

Lu, W., & Hustrulid, W. (2003). A further study on the mechanism of airdecking. *Fragblast*, *7*(4), 231–255. https://doi.org/10.1076/frag.7.4.231.23532

Lundborg, N. (1974). *The hazards of flyrock in rock blasting* (SweDeFo Report DS1974). Swedish Detonic Research Foundation.

Mandal, S. K. (1997). Causes of flyrock damages and its remedial measures. Lecture notes of *Course on: Recent advances in blasting techniques in mining and construction projects* (pp. 130–136). Human Resource Development Center, CSIR-Central Institute of Mining and Fuel Research.

Mangum, H. L., & Bryan, V. C. (1987). Advantages of precise millisecond delay detonators. *Journal of Explosives Engineering*, *4*(5).

McKenzie, C. (2009). Flyrock range and fragment size prediction. In *Proceedings of the 35th annual conference on explosives and blasting technique* (Vol. 2, p. 2). http://docs.isee.org/ISEE/Support/Proceed/General/09GENV2/09v206g.pdf

Mishra, A. K., & Mallick, D. K. (2013). Analysis of blasting related accidents with emphasis on flyrock and its mitigation in surface mines. In *Rock fragmentation by blasting, FRAGBLAST 10—proceedings of the 10th international symposium on rock fragmentation by blasting* (pp. 555–561). CRC Press/Balkem. https://doi.org/10.1201/b13759-78

Mishra, A. K., Nigam, Y. K., & Singh, D. R. (2017). Controlled blasting in a limestone mine using electronic detonators: A case study. *Journal of the Geological Society of India*, *89*(1), 87–90. https://doi.org/10.1007/s12594-017-0563-5

Modarres, M. (2006). *Risk analysis in engineering: Techniques, tools, and trends* (1st ed.). CRC Press/Balkema.

Nieble, C. M., & Penteado, J. A. (2016). Risk management: Blasting rock near concrete inside a subway station in a densely populated urban environment. *ISRM International Symposium—EUROCK 2016*, *2*, 1269–1273. https://doi.org/10.1201/9781315388502-222

Opeyemi, D. (2009). Vibration and flyrock control through better predictions. *Proceedings of the International Society of Explosives Engineers*, 11p.

Pradhan, G. K. (1996). *Explosives and blasting techniques*. Mintech Publications.

Raina, A. K., Agarwal, A., Singh, R. B., & Choudhury, P. B. (2015). Electronic detonators: The psychological edge. *Journal of Mines, Metals and Fuels*, *63*(4), 88–96.

Raina, A. K., & Bhatawdekar, R. M. (2022). Blast-induced flyrock: Risk evaluation and management. In *Risk, reliability and sustainable remediation in the field of civil and environmental engineering* (pp. 209–247). Elsevier.

Raina, A. K., Chakraborty, A. K., Choudhury, P. B., & Sinha, A. (2011). Flyrock danger zone demarcation in opencast mines: A risk based approach. *Bulletin of Engineering Geology and the Environment*, *70*(1), 163–172. https://doi.org/10.1007/s10064-010-0298-7

Raina, A. K., Chakraborty, A. K., Ramulu, M., & Choudhury, P. B. (2006). Design of factor of safety based criterion for control of flyrock/throw and optimum fragmentation. *Journal of the Institution of Engineers*, *87*, 13–17.

Raina, A. K., Wankhende, P., & Singh, P. K. (2016). Controlled blasting for safe excavation of a portion of irrigation canal in close vicinity of a village. *Recent Advances in Rock Engineering*, 82–84. https://doi.org/10.2991/rare-16.2016.13

Rathore, S. S., & Jain, E. S. C. (2007). Studies on flyrock at soapstone quarry for safe working. In P. Moser (Ed.), *Proceedings of the Vienna conference* (pp. 433–438). European Federation of Explosives Engineers.

Rehak, T. R., Bajpayee, T. S., Mowrey, G. L., & Ingram, D. K. (2001). Flyrock issues in blasting. *Proceedings of the Annual Conference on Explosives and Blasting Technique*, *1*, 165–175.

Richards, A., & Moore, A. (2004). Flyrock control—by chance or design. *Proceedings of the 30th annual conference on explosives and blasting technique*, *6*, 335–348.

Rossmanith, H. P. (2002). The use of Lagrange diagrams in precise initiation blasting. Part I: Two interacting blastholes. *Fragblast*, *6*(1), 104–136. https://doi.org/10.1076/frag.6.1.104.8854

Roth, J. A. (1979). *A model for the determination of flyrock range as a function of shot conditions* (Report No. PB81222358). U.S. Dept. of Commerce, NTIS, 61p.

Rustan, A., Cunningham, C. V. B., Fourney, W., & Spathis, A. (2011). Mining and rock construction technology desk reference. In A. Rustan (Ed.), *Mining and rock construction technology desk reference*. CRC Press. https://doi.org/10.1201/b10543

Schneider, L. C. (1996). Flyrock part 1: Safety and causes. *Journal of Explosives Engineering*, *13*(9), 18–20.

Schneider, L. C. (1997). Flyrock-part 2: Prevention. *Journal of Explosives Engineering*, *14*(1), 1–4.

Seccatore, J., Origliasso, C., & De Tomi, G. (2013). Assessing a risk analysis methodology for rock blasting operations. In *Blasting in mines—new trends: Workshop hosted by FRAGBLAST 10—the 10th international symposium on rock fragmentation by blasting* (pp. 51–60). CRC Press/Balkema. https://doi.org/10.1201/b13739-11

Sen, G., & Hayward, M. (2007). Blasting operations: Training strategies. In P. Moser (Ed.), *Vienna conference proceedings* (pp. 395–400). European Federation of Explosives Engineers. https://doi.org/ISBN 978-0-9550290-1-1

Sharma, S. K., & Rai, P. (2016). Assessment of blasting performance using electronic vis-à-vis shock tube detonators in strong garnet biotite sillimanite gneiss formations. *Journal of the Institution of Engineers (India): Series D*, *97*(1), 87–97. https://doi.org/10.1007/s40033-015-0078-4

Shea, C. W., & Clark, D. (1998). Avoiding tragedy: Lessons to be learned from a flyrock fatality. *Coal Age*, *130*(2), 51–54. www.osti.gov/etdeweb/biblio/621062

Singh, C. P., Agrawal, H., & Mishra, A. K. (2021). Frequency channeling: A concept to increase the frequency and control the PPV of blast-induced ground vibration waves in multi-hole blast in a surface mine. *Bulletin of Engineering Geology and the Environment*, *80*(10), 8009–8019. https://doi.org/10.1007/s10064-021-02400-5

Singh, C. P., Agrawal, H., Mishra, A. K., & Singh, P. K. (2019). Reducing environmental hazards of blasting using electronic detonators in a large opencast coal project—a case study. *Journal of Mines, Metals and Fuels*, *67*(7), 345–350.

Singh, S. P. (1995). Mechanism of cut blasting. *Transactions—Institution of Mining & Metallurgy, Section A*, *104*(September–December), A134–A138. https://doi.org/10.1016/0148-9062(96)85138-2

Sjöberg, J., Schill, M., Hilding, D., Yi, C., Nyberg, U., & Johansson, D. (2012). Computer simulations of blasting with precise initiation. In *Proceedings of the ISRM international symposium—EUROCK 2012*. International Society of Rock Mechanics.

St. George, J. D., & Gibson, M. F. L. (2001). Estimation of flyrock travel distances: A probabilistic approach. In *EXPLO conference* (pp. 409–415). AusIMM.

Stojadinović, S., Pantović, R., & Žikić, M. (2011). Prediction of flyrock trajectories for forensic applications using ballistic flight equations. *International Journal of Rock Mechanics and Mining Sciences*, *48*(7), 1086–1094. https://doi.org/10.1016/j.ijrmms.2011.07.004

Thote, N. R., Dhekne, P. Y., & Dongre, R. R. (2013). An overview of accidents due to flyrock: Predictions and mitigative measures. In *Proceedings of the national conference on challenges in 21st century mining—environment and allied issues* (pp. 167–172). Mining Engineers Association of India.

Venkatesh, H. S., Gopinath, G., Balachander, R., Theresraj, A. I., & Vamshidhar, K. (2013). Controlled blasting for a metro rail project in an urban environment. In *Rock fragmentation by blasting, FRAGBLAST 10—proceedings of the 10th international symposium on rock fragmentation by blasting* (pp. 793–801). CRC Press/Balkema. https://doi.org/10.1201/b13759-114

Verakis, H. (2011). Floyrock: A continuing blast safety threat. *Journal of Explosives Engineering*, *28*(4), 32–37.

Verakis, H, & Lobb, T. (2007). 2007G Volume 1—flyrock revisited: An ever-present danger in mine blasting. In *33rd annual conference on explosives and blasting technique* (Vol. 1, pp. 1–10). http://docs.isee.org/ISEE/Support/Proceed/General/07GENV1/07v109g.pdf.

Workman, J. L., & Calder, P. N. (1994). Flyrock prediction and control in surface mine blasting. *Proceedings of the conference on explosives and blasting technique* (pp. 59–74). International Society of Explosive Engineers.

Wu, L., Yu, D., & Duan, W. (2012). The dynamic stress characteristics of air-decked bench blasting under soft interlayer. *Applied Mechanics and Materials*, *101–102*, 400–404. https://doi.org/10.4028/www.scientific.net/AMM.101-102.400

Wu, L., Yu, D., Duan, W., & Zhong, D. (2012). Rock failure mechanism of air-decked smooth blasting under soft interlayer. *Advanced Materials Research*, *402*, 617–621. https://doi.org/10.4028/www.scientific.net/AMR.402.617

Yi, C., Johansson, D., Nyberg, U., & Beyglou, A. (2016). Stress wave interaction between two adjacent blast holes. *Rock Mechanics and Rock Engineering*, *49*(5), 1803–1812. https://doi.org/10.1007/s00603-015-0876-x

Zarei, M., Shahabi, R. S., Hadei, M. R., & Louei, M. Y. (2022). The use of air decking techniques for improving surface mine blasting. *Arabian Journal of Geosciences*, *15*(19), 1–12.

Zhou, Z., Xibing, L. I., Xiling, L. I. U., & Guoxiang, W. A. N. (2002). Safety evaluation of blasting flyrock risk with FTA method. *Safety*, *1*, 1184–1187.

Zijun, W. (2002). Explanation and discussion on flying stones accident in blasting operation. In *The 7th international symposium on rock fragmentation by blasting, Fragblast 7* (pp. 672–675). Metallurgical Industry Press.

7 Flyrock Control Document

A comprehensive account of various facets of flyrock, its impact, and predictive regimes have been discussed throughout the book. However, a quick reference for people who are faced daily with the question of flyrock was thought necessary. Accordingly, this chapter summarizes the flyrock definitions and causes, meaning of blast danger zone, risk assessment and risk management, along with control measures. In summary, the objectives of the chapter are as follows:

1. Knowledge sharing with the industry on the subject
2. Create awareness about flyrock to enhance mine safety and accident prevention
3. Mitigation measures of flyrock
4. Create awareness about safety measures to be observed during blasting to control flyrock

7.1 WHAT IS A FLYROCK?

There is an ambiguity in the definition of flyrock and its subcomponents as brought out in Section 2.1 and further illustrated in Chapter 4, wherein an attempt has been made to clear the confusion in the definitions and terminology. Further, stress has been laid on several aspects of reporting and use of internationally recommended terminology (Rustan et al., 2011).

Flyrock is a rock fragment with a shape (flyrock shape), size (flyrock size), and mass (flyrock mass) that ejects from a blast face to greater distances, and is a result of

- an excessive build-up of explosive gases than the requirement of a blast, arising from excess explosive per hole;
- presence of paths of least resistance for the explosive gases in a bench to be blasted, which include rockmass defects or anomalies (weak beds in competent rock, voids, fractures, and fissure) and poorly designed blast, less stemming length, poor stemming material, poor stemming method, uneven burden, fractured top and/or front face of the bench, delay error or delay malfunction; and
- poorly executed blast (poor blasting practice).

In aforementioned case(s), the high-pressure detonation gases produced in a blasthole focus towards the path of least resistance about the blasthole and lead to flyrock. Difficult mining conditions explained in Section 5.1 have higher propensity of flyrock, and hence the ensuing risk. The aforementioned definition of flyrock caters to many causes of flyrock also that are summarized further in Table 7.1 where ranking of the causes has been assigned as per the number of citations in published literature.

DOI: 10.1201/9781003327653-7

TABLE 7.1
Major Causes of Flyrock as Reported by Several Authors

Rank	Cause	Relation with Flyrock
1	Geological anomalies	Severe flyrock can move in any direction or as dictated by the anomaly
2	Insufficient burden	Flyrock mostly towards front face
3	Insufficient stemming (length)	Cratering, flyrock can be launched at most favourable angle, and can travel to greater distances
4	Inadequate delay	Inadequate relief to a hole or row, possible cratering effect
5	Improper blast layout	Flyrock can move in any direction and distance
6	Excessive specific charge	Multiple flyrocks can travel greater distances, cratering effect
7	Poor quality of stemming	Cratering and hence flyrock towards the top of the blasthole
8	Drilling inaccuracy	If not taken into consideration while explosive charging can be a typical case for flyrock generation
9	Poor confinement	Related to burden and spacing and hence can result in flyrock towards front of the blast face in the case of low confinement or towards the top if over-confined
10	Improper loading of explosives	Not much researched, it has several constituents like density, VoD, watery holes, improper gassing in the case of emulsion explosive, or can be related to over- or under-charging
11	Misfire	If a blasthole fails to fire, it can result in over-confinement of rest of the blastholes and hence cratering and flyrock from the top of the blasthole. Also, wrong handling of misfires can produce flyrock due to several reasons
12	Spacing	Not professionally researched, excessive spacing with corresponding burden may lead to mushrooming of the blasthole towards the top. However, if over-charged, it can result in cratering
13	Density of explosive	Since density has a direct relationship with the VoD of explosive and resultant pressures, wrong explosive selection can be a concern. However, it is not considered to be a major cause of flyrock
14	Waterlogged blasthole	Two issues crop up when charging waterlogged blastholes: (1) if the density of explosive is less than 1000 kg/m^3, then the explosive will float and can be a case of improper loading; (2) mixing of clay or drill cuttings or stemming material with the water produces slush that hampers proper stemming, and the case then shifts to the categories of stemming length and type
15	Choked blast	This is a case of improper relief as any previously blasted muck lying on the toe or completely choking a blast face can lead to cratering or flyrock from the top of the face

There are several other causes of flyrock that have been elaborated in Chapter 3 and should be consulted for proper evaluation of the blast and rockmass conditions.

7.2 WHO ARE THE SUBJECTS OF FLYROCK?

Two categories of subjects of flyrock have been discussed in Section 5.2 along with the consequence and relative cost of flyrock, viz. belonging and not belonging to the owner of the mine. Such objects, referred to as OCs, include habitats nearby along with humans and their assets of any nature, as provided in Table 5.1. A mine owner, a blasting engineer, planner, and even legal authorities should have a complete knowledge of such subjects. All such objects are subject to flyrock and can have severe consequences, if hit by one. Such accidents can lead to penalties and extreme actions by the legal authorities. It must be remembered that *flyrock can* be fatal.

7.3 WHAT IS BLAST DANGER ZONE?

Blast danger zone (BDZ) or blast security area is a zone around the blast area, which needs to be cleared of people or any mobile assets before the blast is conducted. The immobile asset will need to be protected from flyrock. Many standards recommend a BDZ of one half of the distance of the OCs from the blast site, while most of the standards are purely event based and lack scientific basis, as can be verified from Figure 7.1.

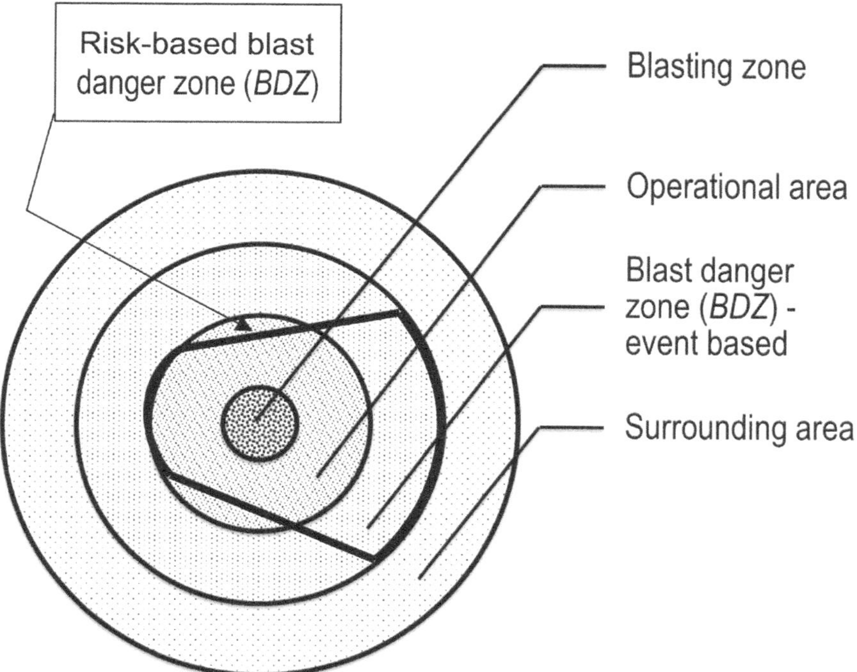

FIGURE 7.1 Difference between event-based and scientific blast danger zone.

The effort by the mine management should be focused on optimization of the BDZ through different means, as discussed earlier and summarized further.

7.4 CAN FLYROCK BE PREDICTED?

Flyrock is an uncertain phenomenon in comparison to ground vibrations, air overpressure, fragmentation, and throw during blasting. There are two aspects to flyrock prediction. One is whether flyrock will occur or not, and it is essential for all mines facing such problems to ascertain such probabilities. Methods to estimate the flyrock probability, *P(E)*, have been discussed in Section 5.5 and must be used to determine these as these will be specific to individual sites.

The second aspect is the flyrock distance prediction that can be attempted with a fair degree of accuracy using different site-specific methods like empirical, multivariate analysis, and artificial intelligence tools. Many methods mentioned in Chapter 4 can be invoked to predict the flyrock distance. However, it is good to start with the maximum distance criterion (Lundborg, 1974) as given here:

$$R_f = 40 \; to \; 260 \; d^{\frac{2}{3}}$$

where R_f is the flyrock distance in m, ρ_r is the density of rock in kg/m^3, d is the diameter of the drill hole in inches, and k_x is the size of rock fragment in m travelling maximum distance.

It is good to start with maximum distance, i.e. factor 260 in the aforementioned equation, and arrive at the factor for the mines after sufficient experimentation. If this does not work for the mines, try changing the drill diameter to restrict the flyrock distance.

The methods discussed in Chapter 4 can also be tried. Specifically, one provided in Equation 4.41 should be tried and evaluated to match the blast design. The correction factor and its method of application will need site-specific calibration for determination of flyrock distance.

7.5 FLYROCK RISK

The flyrock probabilities, *P*(*E*), must be established in a particular mining condition. It is important to keep a record of throw, all incidences of flyrock along with its direction, size, shape, mass, and number of blasts that produced flyrock along with their distance of travel, wind velocity, and wind direction (relative to blast face direction or flyrock direction). The cost of a flyrock event, *C*(*E*), can be established through different methods, including known regulatory penalties or, if possible, through insurance agencies. Extreme situations of fatality of persons not belonging to the mines should also be considered while evaluating such costs, including the cost of closure of mines due to implementation of regulations in case of a flyrock event. Once the two constituents of risk are known, the risk due to flyrock can be evaluated as follows:

$$Risk = P(E) \times C(E)$$

where E denotes a flyrock event.

Once risk is evaluated, proper risk management protocols should be put in place. This method can be used to define a scientific BDZ.

If $C(E)$ is not known or cannot be worked out, the "threat ratio" method in Section 6.2.5 can be invoked to determine the risk level and fix the BDZ. Normalized values of $C(E)$, i.e. ratio of maximum penalty to a particular penalty, can replace the abscissa of threat ratio, if actual costs of the event are available.

7.5.1 Risk Classification

The managers, planners, and blasting engineers must evolve a risk matrix for flyrock. Although Figure 6.2 provides a basis for such classification with a broad risk matrix in Table 5.4, an example of devising such matrix is defined in Table 7.2.

It must be borne in mind that Table 7.2 is for demonstration purpose only and every mine must develop their own risk matrix to arrive at tangible solutions for flyrock risk management. However, it can be observed from Table 7.2 that there is a scope for buy-in with the risk, as marked by stepped lines/shaded areas, and that a mine owner or engineer or even legislators/legal authorities must exploit in the interest of the mines and ore exploitation.

7.5.2 Risk Management

The basic premise of risk management in blasting (Revey, 2000a, 2000b) defines risk management system that has four main tenets, viz. practical and adequate blast design, pre-qualification (experienced staff), specifications (rules defining performance and safety protocols), and oversight (implementation as planned by the management).

Risk management includes the process of decision-making involving the following:

1. Avoiding the risk means taking best possible preventive measures to ensure that no flyrock occurs. This may include risk classes with minimal risk levels defined in Table 7.2.
2. Share risk through insurance cover (may not be applicable or available in many places). This will be in tune with the risk classes in Table 7.2 that show minor to moderate risk level.
3. Reduce risk through control measures. When preventive measures fail to achieve a particular risk level, the control measures will need to be put in place. This may mean reducing the BDZ or threat level. In extreme cases, the system changes like blasthole diameter reduction can be tried.
4. Accept and own the risk through preventive, control, and detection measures. If the management is conscious and implementing all measures, it can opt to accept the risk. However, in such a case, the complete onus will be on the mine management and any flyrock event and consequences will have to be taken care of.

However, it must be understood, that safety is a continuous process and any deviations from the statutory and management policies with respect to flyrock can result

TABLE 7.2
An Example of Development of a Risk Matrix (the Cost Values Are Assumed and Not Actual)

Risk Matrix		Event Classification	Minor	Moderate	Moderately High	High	Very High	Catastrophic/Major
			Damage	Damage	Damage	Damage	Damage	Damage
			Injuries	Injuries	Injuries	Injuries	impairment	Fatality
Probability/ Likelihood	1.0	Certain	20,000	50,000	100,000	250,000	500,000	1,000,000
	10^{-1}	Very likely	2,000	5,000	10,000	25,000	50,000	100,000
	10^{-2}	Likely	200	500	1,000	2,500	5,000	10,000
	10^{-3}	Very possible	20	50	100	250	500	1,000
	10^{-4}	Possible	2	5	10	25	50	100
	10^{-5}	Not likely	0	1	1	3	5	10
	10^{-6}	Very unlikely	0	0	0	0	1	1
Cost description		Within mine						
		Outside mine	Minimum	Low	Moderate	High	Very high	Maximum (mine closure, fatality)
Assumed cost of an event (example and not absolute)			20,000	50,000	100,000	250,000	500,000	1,000,000
Threat ratio			0.1	0.2	0.4	0.6	0.8	1

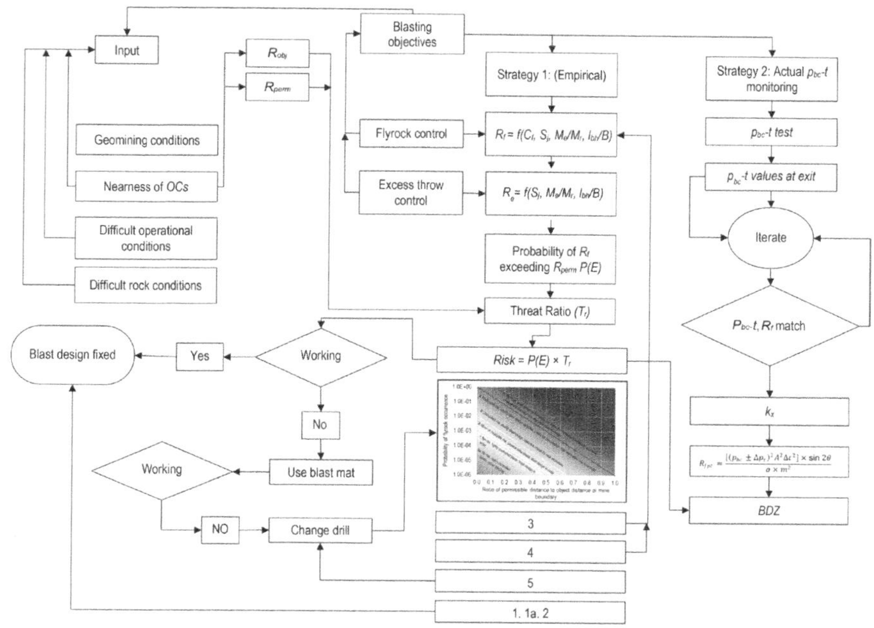

FIGURE 7.2 A comprehensive strategy, including risk identification and remedial actions, for flyrock prevention and control.

Source: Modified after Raina (2014).

in serious consequence. The mine closure is a worst-case example. Resumption of blasting and confidence-building measures can be quite costly.

A comprehensive methodology for flyrock prevention and control has been evolved and is presented in Figure 7.2. All procedures for deploying such strategy have been discussed at various places in this document.

7.6 FLYROCK PREVENTION AND CONTROL

The methods of flyrock prevention (Section 6.2) and control (Section 6.3) are summarized here in order of the significance of causes. Specific measures are compiled in Tables 7.3 and 7.4.

1. Identification of geological anomalies in advance and adjusting blast design, accordingly, taking pre-emptive measures to prevent flyrock. This includes a complete know-how of the rock formation, its characteristics, and presence of incompetent rock layers within the bench particularly dipping into the blast face. Communicate the information to drillers and blasters about such conditions, and apply proper measures for preventing flyrock.

TABLE 7.3
Measures to Be Taken for Prevention of Flyrock

Activity	Components	Preventive Measures
Survey	Geology	• Survey of geology • Knowledge and mapping of ◦ Local geology/lithology ◦ Rock anomalies like voids, fissures, cracks, joints, mud seams
	Pre-drilling inspection	• Pre-drill inspection—blast face specific: faults, backbreak, irregularities, conduct laser profiling if available or use other measurement methods • Record any indication of voids while drilling • Clear the top surface of the bench before the drilling starts
	Pre-loading inspection	• Inspect blastholes before loading; log burden, spacing, hole depth, bench height, and presence of water • Pay special attention to front row hole(s) drilled pattern • Make a loading sheet for filling blastholes with explosive as required; use in-hole deck if charge exceeds than the calculated
	Post-firing inspection	• Post firing inspection—misfire, cut-off charge, bootleg, backbreak
Blast design	Stemming length	• Adequate stemming length (not less than 0.7 times of the burden, higher values for difficult geomining conditions • Fill blastholes up to top of the stemming zone collar
	Stemming material	• Use of proper stemming material, avoid drill cuttings if these are of fine nature • Stemming material size >6 mm or hole diameter/20 size (must for watery holes) • Stemming material should be free from the bigger size fragments • Try airbags or stemming plugs
	Burden	• Define blastability of rockmass • Burden should be in tune with the drill diameter based on blastability • Eliminate too low or too high burden • If large toe burden exists, it should be blasted prior to the main blast • First row burden to be cross-checked, if crest is broken, increase stemming length
	Drilling accuracy	• Drill pattern should match the design blast pattern • Drilling deviation should be minimum, check specifically if length of blastholes is more than 15 m • Use inclined holes if possible and if toe formation is frequent
	Subdrilling	• Decide on subdrilling whether required or not (formation dependent), if yes follow standard procedure to calculate the depth (normally 0.3 × Burden), avoid if soft seam is at the bottom of bench
	Bench height	• Match bench height to burden ratio (≥3) • Benches of 10–18 m are considered the most economical and least hazardous to work

Activity	Components	Preventive Measures
	Delay	• Establish delay scatter and reliability, conduct occasional random checks for delay accuracy • Delay sequence/pattern to match the dynamic properties and design • For multiple rows, longer delays between later rows • Evaluate the delay timing to ensure the appropriate firing sequence • Try electronic detonators
	Initiation method	• Use bottomhole initiation and avoid use of detonating cord • Follow the initiation sequence devised and that gives best results
	Blasthole diameter	• Blasthole diameter should be chosen according to the distances up to which flyrock can be tolerated use criterion for maximum flyrock distance, size in relation with the OCs, and blasting permissions/regulations • Deploy optimum charge diameter
	Explosive diameter	• For difficult geomining conditions, use lesser diameter charges
	Explosive type	• Match the impedance of explosive with the rock. Avoid using high gas component explosive in formations with greater propensity of flyrock

2. Stemming control
    i. Stemming length should be equal or more than 0.65 × Burden. In critical conditions, the stemming length can be equal to burden.
    ii. Stemming material should be stone chips of not more than *d*/20 size (*d* is the drill diameter in $\times 10^{-3}$ m. No fragments of larger size should be used in stemming to prevent ejection and damage of firing connections during tamping.
    iii. Stemming should be well-tamped.
    iv. Test the use of stemming plugs and document any improvements.
3. Check for highly broken blasthole collar and take measures to prevent flyrock ejection.
4. Check for highly indented face conditions resulting in least burden. Take measures not to charge or use decking in this area of bench.
5. Remove any rock fragments from the blasthole collar.
6. Check and test electronic detonators for their efficacy in controlling flyrock.

7.6.1 General Rules

1. Clear the area near the blast of machines and within BDZ from unauthorized personnel.
2. Communicate all relevant information—meaning interaction between survey, planning, and blasting departments.
3. Blasting crew to stay in a blasting shelter during firing.

TABLE 7.4
Measures to Be Taken for Control of Flyrock

Activity	Components	Control Measures
Blast danger zone	Hazard assessment	• Make a complete hazard assessment before conducting a blast
	Safety rules	• Define safety rule or statement, publish, and disseminate—buy in with blasting crew
	Area security	• Statutory regulation of national security blasting should be strictly obeyed and conveyed to concerned department along with possible penalty for failure to observe • Definition of BDZ in clear terms • Control and exclusion of the BDZ and guarding of the safety zone • Crew to clear the area before blasting or take shelter • Shot firer to use shelter (specially designed or recommended) invariably • Guarding and restricting of blast area during loading with guarded entrances • Guards to enforce proper security, including non-entry of prohibited items
Drilling and blasting	Drilling	• Experienced and fully trained drilling/blasting crew is essential • Drillers to report voids while drilling, and monitor changes in drilling speed • Modern-day drills to be deployed and drill data scrutinized before developing a charge configuration
	Blasthole loading method	• Proper drilling and charging of the first row of holes are important • Blastholes to be filled with explosive that is distributed along its length • Abandoned blastholes should be backfilled • Inspect the blast bench for unevenness, projections, fractures, incompetent strata in competent rock, toe burden, backbreak, jointing, mud seams, voids, and zones of weakness and take measures to adjust charge configurations
Total flyrock control	Special method	• Muffling, covering using rubber mats, sandbags, industrial felt, conveyor belts used tyres with sandbags, tarpaulin, closed meshed mats
Communication and training	Communication	• Establish clear lines communication and supervision of the blasting activity • Effective communication between driller and blasting engineer
	Training and education	• Frequent training of blasting and drilling crew in blast loading, initiation methods—trainings should be focused on safety in blasting and flyrock
Post-blast	Post-firing inspection	• Inspection of the blast area by shotfirer for misfires, backbreak, etc. who if satisfied should intimate engineer-in-charge to declare the success of blast and clear BDZ restriction. Use stipulated protocol for handling of misfires

4. Provide sufficient warning signals and displays at critical locations.
5. Alarm before firing the shot.
6. Run awareness campaigns and conduct trainings in line with safety plan.
7. Document and analyse the reasons for flyrock—take proactive measures to prevent such occurrences in future.

7.7 BLASTING ENGINEERS CHECK SHEET FOR PREVENTION OF FLYROCK ACCIDENTS

A checklist of information for blasters and engineers that may be of use to other concerned persons is provided in Table 7.5. The mine management should take steps to examine and finalize their own checklist of activities, prevention, and control measures for flyrock in line with the one provided here.

7.8 A WORD TO REGULATORS AND RESEARCHERS

One of the issues faced by people concerned with flyrock is non-reporting of flyrock. It is felt that it is not intentional. Since there are serious consequences of flyrock,

TABLE 7.5
Check Sheet for Flyrock Prevention and Control

S. No.	Measure/Activity	Value If Applicable	Response
1	Are there any weak areas in the stemming zone?		
2	Is the burden at stemming zone too less than the burden at toe?		
3	Is burden of any hole too less or too much than designed?		
4	Are there any incompetent strata in competent rockmass		
5	Are there any stray rocks on the blastholes?		
6	Is stemming length proper?		
7	Is stemming material proper?		
8	Have you checked the surface connections?		
9	Have you shifted equipment to safe distance?		
10	Have you shifted blasting crew to a safe area?		
11	Blast danger zone (BDZ) distance?		
12	If blast is very near to locality, did you use blast mat?		
13	Are there any public utilities within the BDZ?		
14	Are there any visitors within the BDZ?		
15	Have you secured the BDZ?		
16	Have you protected yourself and the shotfirer?		
17	Have you issued initial alarm?		
18	Have you issued final alarm?		
19	Record of flyrock information		

mines shy away from reporting. This is to the disadvantage of the planners, mine management, legal authorities, and researchers equally. A methodology to discriminate between the throw and flyrock has been described in detail in this work. The method is intended to discriminate between recordable and reportable flyrock. If the criterion is adopted, it will enable the blasting engineers to record the flyrock and report also. Creation of a database on flyrock can be quite useful to the international community for its well-being and at the same time help to evolve rules and guidelines, based on science and logic, to have a complete control over the occurrence of flyrock.

Scientists and scholars can take advantage of the issues raised herein and analyse their reports of flyrock while using standard terminology and nomenclature, publishing flyrock data in proper perspective, i.e. flyrock is uncertain, and its study must incorporate both probabilities and predictive models together. If cost of a flyrock or consequence is known that must be documented accordingly. Mere use of intelligent methods for flyrock distance prediction, without known probabilities and risk included, precludes the chances of a holistic solution to flyrock issue for the benefit of the science, mining fraternity, and the public.

REFERENCES

Lundborg, N. (1974). *The hazards of flyrock in rock blasting* (SweDeFo Report No. DS1974). Swedish Detonic Research Foundation.

Raina, A. K. (2014). *Modelling the flyrock in opencast blasting under difficult geomining conditions*. Ph.D. Thesis, Indian Institute of Technology—ISM. www.iitism.ac.in/pdfs/departments/mining/Research-degrees-completed.pdf

Revey, G. (2000a). Evaluating and managing blasting risk part I. *The Journal of Explosives Engineering*, 6–13.

Revey, G. (2000b). Evaluating and managing blasting risk part II. *The Journal of Explosives Engineering*, 14–19.

Rustan, A., Cunningham, C. V. B., Fourney, W., & Spathis, A. (2011). Mining and rock construction technology desk reference. In A. Rustan (Ed.), *Mining and rock construction technology desk reference*. CRC Press/Balkema. https://doi.org/10.1201/b10543

Index

D

E

F

G

H

I

K

L

M

O

P

R

S

T